THE USE OF FUNGI AS FOOD AND IN FOOD PROCESSING,

Part II

Author:

WILLIAM D. GRAY
Adjunct Professor of Biological Sciences
Northern Illinois University
DeKalb, Illinois

CRC Press
Taylor & Francis Group
Boca Raton London New York

CRC Press is an imprint of the
Taylor & Francis Group, an **informa** business

First published 1973 by CRC Press
Taylor & Francis Group
6000 Broken Sound Parkway NW, Suite 300
Boca Raton, FL 33487-2742

Reissued 2018 by CRC Press

A Library of Congress record exists under LC control number: 70576630

Publisher's Note
The publisher has gone to great lengths to ensure the quality of this reprint but points out that some imperfections in the original copies may be apparent.

Disclaimer
The publisher has made every effort to trace copyright holders and welcomes correspondence from those they have been unable to contact.

ISBN 13: 978-1-315-89838-4 (hbk)
ISBN 13: 978-1-351-07748-4 (ebk)

Visit the Taylor & Francis Web site at http://www.taylorandfrancis.com and the
CRC Press Web site at http://www.crcpress.com

THE AUTHOR

William D. Gray is Adjunct Professor of Biological Sciences, Northern Illinois University, DeKalb, Illinois. He has retired from active teaching and presently resides in Lancaster, Ohio.

Dr. Gray received his A.B. degree from DePauw University in 1933 where he received the Distinguished Alumnus Award in 1965. His Ph.D. degree was awarded by the University of Pennsylvania in 1938 when he was also named a National Research Fellow at the University of Wisconsin.

Dr. Gray has authored four books and more than 70 research reports in various journals.

This book originally appeared as part of an article in *CRC Critical Reviews in Food Technology*, a quarterly journal published by The Chemical Rubber Co. We would like to acknowledge the editorial assistance received by the Journal's editor, Thomas E. Furia, CIBA-GEIGY Corp. Dr. C. M. Christensen, Professor of Plant Pathology, University of Minnesota, served as referee for this article.

PREFACE

When the opportunity to prepare a paper on fungi as food first arose, I attempted to describe briefly the methods of producing mushrooms on a commercial basis, the ways in which fungi are used (principally in the Orient) to process certain foods, and the vast, relatively unexplored, potential of fungi as sources of much-needed edible protein. In connection with this, certain toxic molds were also considered. Although this paper elicited rather pleasing responses, I felt that it was incomplete and so this second part represents an attempt at completion – an impossible task as long as research in applied mycology continues.

Obviously, the people of this world cannot depend upon the collection of wild fungi as a major food source, but neither could they depend solely upon cane sugar, which latter fact does not prevent the use of millions of acres of land annually for the production of this carbohydrate. Both food items play their roles in man's total food supply – it just happens that one is exploited to a far greater extent than the other. During the 40-odd years that I have been involved in considering the relation of fungi to human affairs I have encountered hundreds of laymen who have expressed a desire to be able to collect and safely eat wild mushrooms. These "laymen" ranged from laborers to highly trained professional people. There is no great mystery involved in the identification of fungi and what they needed most was encouragement. Hence, no attempt is made here to change the layman into a taxonomic mycologist – on the contrary he is merely provided with encouragement and it is pointed out to him that with the expenditure of a small amount of effort he can learn to recognize many edible fungi. However, the eating of wild mushrooms by the uninitiated carries with it the danger of getting one of the few poisonous species. Therefore, a section on poisonous mushrooms and mushroom poisoning has been inserted to provide the beginner with advance knowledge of the possible consequences of his not having learned his lesson well.

Most laymen know about LSD and peyote but only a very small number are aware that hallucinogenic mushrooms exist. Such fungi are certainly not ingested for their food value, but the logical place to include them in a review article seems to be in connection with a consideration of fungi as food. Since ethnomycology is scarcely two decades old, the present discussion is just a beginning, and further interesting information on the role of hallucinogenic fungi in man's cultural and religious heritage may confidently be expected to be brought to light.

Few laymen are aware that cookbooks devoted solely to the preparation of mushrooms exist. This fact is brought to their attention and a few recipes are presented in detail. No attempt is made to exhaust the subject but as in the remainder of the book, enough is presented to give a view of the potential and the reader is directed to the relevant literature so that he can pursue the subject in greater depth if he chooses.

William D. Gray
Lancaster, Ohio

TABLE OF CONTENTS

INTRODUCTION

In an earlier work which also appeared under the title, The Use of Fungi as Food and in Food Processing, I discussed in some detail those species of fleshy fungi which have been cultivated on a large scale by man for food purposes, the use of fungi in the processing of certain types of cheeses and in the aging of meats, fungus-fermented Oriental foods, the potential of fungi for making vast contributions to the world protein pool, and certain toxic substances produced by mold-type fungi. Although occasional references were made to the use of wild mushrooms as food, in addition to poisonous mushrooms and hallucinogenic mushrooms, these subjects were not discussed in detail. The show of interest and pleasing response to this earlier work, especially from individuals in non-mycological areas, were such that it was felt that a more detailed discussion of these subjects which had been neglected in the first volume would now be of some value.

The present work is not designed for nor pointed toward the professional mycologist, and it is hoped that if it falls into such hands it will be treated with some measure of charity. The specific objective is to introduce this subject to the non-mycologically trained person or to the beginner in the field of mycology so that he may gain some concept of the scope of the subject. It is most emphatically not designed to replace reliable mushroom manuals and several are listed in the section which follows. Authority names for mushroom species have been omitted with some misgivings. Obviously, these names are of considerable importance to the serious taxonomic mycologist, but they are definitely unnecessary for the potential or beginning mycophagist. Also, in the present work there is little or no involvement in matters of disputed nomenclature. Taxonomic mycologists have had differences of opinion for many decades regarding such matters as the validity of the specific names applied to certain fungi, whether or not a species should be placed in some other genus, or whether a new genus name should be created. Even today, different mycologists will use different scientific names for the same species of fungus. Hence, many of the scientific names used here are those I have learned – some of them from the authorities of the past. Thus, some of the names used may not be the names used by others. In the matter of name selection, I do not claim to be right; I merely suggest that I may be. In citing the reports of other workers, the scientific names that they employed are used here. These brief remarks should not be interpreted as implying that problems of mycological nomenclature are not important. Such matters are very important in the field of taxonomic mycology, but the general field of mycology must be far broader than taxonomy or it will ultimately stifle itself. I must state, as I have stated many times before, that the true significance of the fungi ultimately rests in what these organisms do and what man can do with them, not what they are called. Hopefully, the many problems of taxonomic mycology eventually will be solved and all mycologists will agree on what constitutes the valid scientific name for each fungus species (although this seems a remote possibility at the moment). Meanwhile, with a maximum of encouragement and a minimum of training (but not too minimum), the average interested individual can readily learn to identify a rather surprising number of delicious, edible, wild mushrooms. The present work is designed to provide the encouragement, help generate the enthusiasm, and supply some of the basic training.

As in the earlier work, the concept of the term mushroom used here may not be shared by all. In the present instance the term is applied to any fleshy fungus (edible or poisonous) with gills (the agarics), as well as any other type of fleshy fungus that is edible. Thus, under this concept the white-fleshed edible puffballs such as *Calvatia gigantea* are mushrooms while the black-centered (and possibly poisonous) *Scleroderma vulgare* is not. The term toadstool is used in only one instance, since this term has no valid meaning; toads rarely if ever sit on mushrooms, although some years ago a zealous photographer for *Life Magazine* apparently managed to persuade a toad to sit on a large mushroom long enough to snap a rather incongruous picture.

Since this paper is primarily concerned with the use of fungi as food, the first section is devoted to a discussion of the collection and identification of some of the common, easily identified, edible wild mushrooms. This section is definitely not designed to supplant a reliable mushroom manual. On the contrary, it is designed primarily to introduce the beginner to the subject, warn him of some of the possible pitfalls, and to point out that the subject is not nearly as difficult as it might appear at first. Other writers might use other species as their examples, but the examples presented here are either those species used by me and/or most commonly recommended by others. For example, I rarely eat the Sulphur Mushroom (*Polyporus sulphureus*), but it is presented here because it is common and because many mycophagists consider it delicious.

Even though it involves a slight amount of repetition, since there was a similar brief section in the first paper of this series, a section concerned with the use of fungi as animal food is again included, because additional information has come to light and it is now beginning to appear that certain animal species may rely more heavily on mushrooms for part of their food supply than was formerly believed. Specific studies are being conducted in this area, and we are no longer solely reliant upon casual observations. Also, it is known that certain wild animals ingest with impunity mushrooms that are definitely poisonous to man. This latter knowledge causes a question to be raised concerning the wisdom of using animal feeding experiments as a means of making reliable determinations of the toxicity of certain mushrooms to man.

A discussion of poisonous mushrooms and various classified types of mushroom poisoning is included because the beginning mycophagist should quickly become well aware of the possible consequences of making a mistake in identification. A list of those mushrooms reported to be poisonous has been compiled, even though many of the species listed have been reported to be poisonous by only one author. Brief descriptions of symptoms associated with the various types of poisoning are presented and, where known, the chemical structures of the active agents are illustrated. Hallucinogenic mushrooms are treated in a section separate from poisonous mushrooms, although it is not clear at this point whether the most fascinating mushroom species of all is both poisonous and hallucinogenic, or merely the latter. Hallucinogens are also included for the purpose of introducing the reader to the new and developing field of *ethnomycology,* of which the founder and most active worker is R.G. Wasson. From his work and the work of others it is beginning to appear that mushrooms may have played a far greater role in man's cultural and religious heritage than was formerly supposed, although it has yet to be definitely established that the Fly Agaric was in reality the only plant ever deified by man or that Judeo-Christianity had its origin in a small cult of mushroom eaters.

The average individual seems almost totally unaware of the myriad ways in which mushrooms may be prepared for eating, and apparently is usually content with a steak smothered in mushrooms or an occasional salad containing a few chopped mushrooms, usually the immature buttons of *Agaricus bisporus.* The facts are that mushrooms may be prepared in a great many ways and served as hors d'oeuvres, soups, main dishes, etc. They have recently become a favorite food of many individuals in the growing cult of weight-watchers, especially those who are compulsive eaters. A surprisingly long list of dishes that include mushrooms has been compiled and presented, although the reader will undoubtedly be able to add to this list when he next reads a newspaper or magazine. No attempt has been made to exhaust the subject but representative recipes of the various categories of mushroom-containing dishes are presented in order to give some idea of the potential of this particular food item, and various books are cited which are devoted solely to the cooking of mushrooms.

THE COLLECTION AND IDENTIFICATION OF EDIBLE WILD MUSHROOMS

When on occasion a non-mycologically trained person becomes interested in mushrooms as a pleasant hobby, as a source of different and unusual types of foods, or merely as an excuse for enjoying the out-of-doors, one of the first disconcerting facts that he is likely to encounter is that there is a tremendous gap between most professional mycologists and the budding enthusiasts, such as himself. Over the past several centuries, a rather imposing number of excellent scientists have contributed to an impressively large body of facts concerning fungi. Unfortunately, only a very small percentage of these excellent scientists seems to have made any sustained attempt to bring the collective and often generally interesting findings of their field to the attention of laymen, at least not in a language that the latter can easily understand.

Therefore, when the layman develops a greater and greater interest in mushrooms, he usually soon finds himself confronted with what appears to be an overwhelmingly great mass of terminology with which he is largely unfamiliar and an even larger mass of literature, most of which is not readily available to him. At this point, only the most dedicated and most deeply interested amateur is likely to continue his quest. I have no wish to lull the beginner into a false sense of security, since hastily or carelessly made identifications can and have led to tragic results. However, I would like to reiterate that learning to recognize a rather considerable number of edible and poisonous species of mushrooms does not present as formidable a problem as might be thought at first. It might be well for the amateur to start with a specimen of a species that can be almost positively identified by means of suitable illustrations and then check his identification by using a key on the same specimen. In this manner he will learn the terminology used in the identification of that species, much of which will also be used in the identification of other species. As additional species are identified, he will become acquainted with additional terms; in this manner he can gradually increase his mycological vocabulary a few terms at a time rather than attempt to learn the meanings of large numbers of new terms at one time.

What should be stated at the outset is that the amateur mushroom enthusiast does not need to become, nor is it likely that he has any wish to become, a professional mycological taxonomist. His primary objective should be to learn to recognize quickly, beyond a question of doubt, some of the common edible species, but more importantly to be able to make positive identifications of the common poisonous species. In fact, learning to identify the poisonous species is probably more important at first than learning to identify the edible ones; furthermore, while this latter capability may often lead to disappointing results if he is looking for mushrooms for the table, it will not lead to tragic or very unpleasant ones. It should again be stated that under no circumstances should any mushroom be eaten until a positive identification of the species has been made: (1) it should resemble the illustration, (2) it should be keyed out with certainty, (3) it should fit the species description in all ways, and (4) it should definitely be listed as an edible species in a reliable manual.

Thus the beginning mushroom enthusiast really has no need for an extensive mycological library. He can start with one or two mushroom manuals that have been written in such style that their use is quickly mastered by the uninitiated. I favor three such manuals, but from these the reader is left to make his own selection. These three are (1) *Common Fleshy Fungi* by Clyde M. Christensen,[37] (2) *The Mushroom Handbook* by Louis C. C. Krieger,[110] and (3) *The Mushroom Hunter's Field Guide* by Alexander H. Smith.[172] In all probability, if one book is purchased and interest continues to grow, the enthusiast will find himself purchasing others, since he will soon find that no single manual will enable him to identify every mushroom that he finds. The three books mentioned above have the advantage that they include many of the most common and conspicuous mushrooms, and for the beginner that should suffice. However, if the enthusiast continues to pursue his hobby seriously, he should by all means become acquainted with Smith's[171] two-volume *Mushrooms in Their Natural Habitats,* the second volume of which consists of a viewer and excellent color transparencies of many of the common fleshy fungi.

Any of the above manuals may well be supplemented with several easily obtained shorter papers, which, while limited in their coverage, often include excellent photographs and descriptions of several of the more commonly encountered fleshy fungi. For example, available upon request from the Canada Department of Agriculture is Mushroom Collecting for Beginners by J. W. Groves.[83] This pamphlet may be obtained by requesting Publication 861 from the Plant Research Institute, Central Experimental Farm, Ottawa, Canada. Similarly, Korf and Burdsall[108] have written Common Mushrooms: Edible and Poisonous, which may be obtained on request as Extension Bulletin 386 of the New York State College of Agriculture at Cornell University, Ithaca, New York. While the pamphlet Edible Wild Mushrooms by Clyde M. Christensen[38] describes only a few species, it has excellent photographs in color which will also be of help to the beginner. It may be obtained on request from the Agricultural Extension Service, University of Minnesota, St. Paul, Minnesota, as Extension Bulletin 357. The extreme usefulness and popularity of this latter pamphlet are exemplified by the fact that by late 1971, 50,000 copies had been distributed.

Most mushroom manuals will give specific details for collecting mushrooms and preparing them for transportation to the laboratory or home. There are a variety of ways in which this may be done, so no one system is especially recommended here. The following points should be observed: (1) as much dirt, rotting wood, dead leaves, etc. as possible should be removed from the bases of the mushrooms in order to have them as clean as possible; (2) specimens of different species should be kept separate – this is especially important for the beginner; (3) the mushrooms should be so packed that they do not become broken or crushed; (4) the entire mushroom should be collected in each instance; sometimes the volva (cup at base) is small and inconspicuous or partially underground and will not be seen if the stipe is broken off at ground level, an oversight that could lead to unfortunate results; (5) they should not be allowed to dry out too quickly. Medium-sized paper bags (not plastic) and a basket or box to carry the bags in when they are filled are usually quite adequate to transport the mushrooms, and a knife should always be carried to cut loose wood-inhabiting species and to dig up soil-dwelling types to be sure that the entire structure is obtained.

Although attempting to learn a large number of terms at one time is to be avoided by the beginner, there is a minimum number with which he must become immediately familiar. Keys and species descriptions are based upon the characteristics of various mushrooms parts (Figure 1), so these must

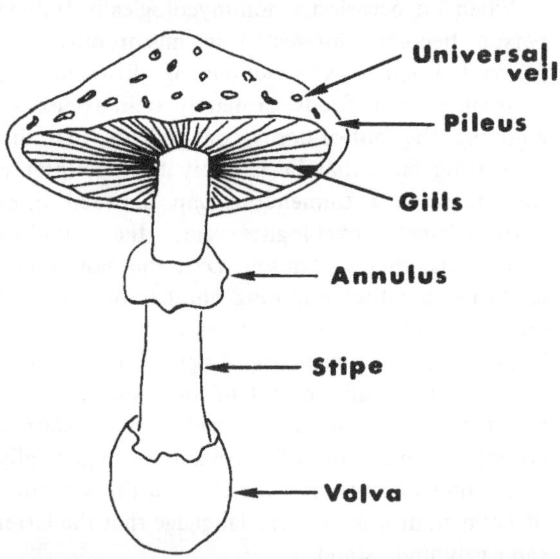

FIGURE 1. Diagram of a gill fungus (agaric) showing the various parts which a mushroom of this type may have.

be known before any identification may be made. In an agaric (gill fungus), these parts are not numerous: *volva* (cup) at the base of the stalk of the mushroom; *stipe* (stem or stalk-line structure); *annulus* (ring of tissue on the stipe); *veil* (fringe of tissue hanging from the margin of the cap) – the gills are covered with a thin layer of tissue and as the cap expands this tissue breaks, forming an annulus on the stipe, a veil at the margin of the cap, both or neither, depending upon how much tissue is involved and where it breaks when the cap expands; *pileus* (cap or expanded upper portion of the mushroom); *gills* (plate-like structures on the lower side of the pileus upon which the microscopic basidiospores are formed); and *universal veil* which covers the entire mushroom when it is young and may appear as large, scattered patches of tissue on the pilei of mature, expanded specimens of species such as *Amanita muscaria* (Figure 2) or is totally missing in others. Some mushrooms may have all of these parts, whereas others do not. Some may have only pileus, gills, and stipe, and some species such as *Crepidotus malachius* (Figure 23), in which the stipe is

FIGURE 2. Mushroom sizes may vary over a very wide range depending upon species. Compare the large basidiocarps of *Amanita muscaria* with the tiny ones of *Marasmius siccus* (insert, upper left). Individual basidiocarps of the same species of *Amanita* may show considerable variation, and the amateur mycophagist should be extremely wary. For example, in the smaller basidiocarp the volva is quite obvious but the annulus is not. However, the larger basidiocarp has a definite annulus but the volva is not so obvious except to the experienced. A careless and cursory identification might place the smaller in the genus *Volvaria* and the larger in the genus *Lepiota*, possibly with unpleasant results. (Photograph of *Marasmius siccus* courtesy of C. M. Christensen.[38])

lacking, consist of only pileus and gills. The presence or absence of some of these parts, their color, size, shape, etc. all assist in making an identification possible.

The pileus is usually the largest and most obvious part of an agaric and its characteristics are especially useful in making an identification. Although most agarics have the general appearance of an umbrella, not all mushroom pilei are shaped alike, and nine general shapes are recognized, as diagrammed in Figure 3. Thus, the campanulate type of pileus characteristic of the Shaggy Mane, *Coprinus comatus,* shown in Figures 4 and 30, could not be easily confused with the infundibuliform (funnel-shaped) pileus of the Chantrelle, *Cantherellus cibarius,* which is shown in Figure 25. Whether or not the surface of the pileus is smooth, hairy, or scaly, its color, whether or not it is sticky, and whether or not it may be peeled easily are all characteristics which are useful and often necessary in making a positive identification.

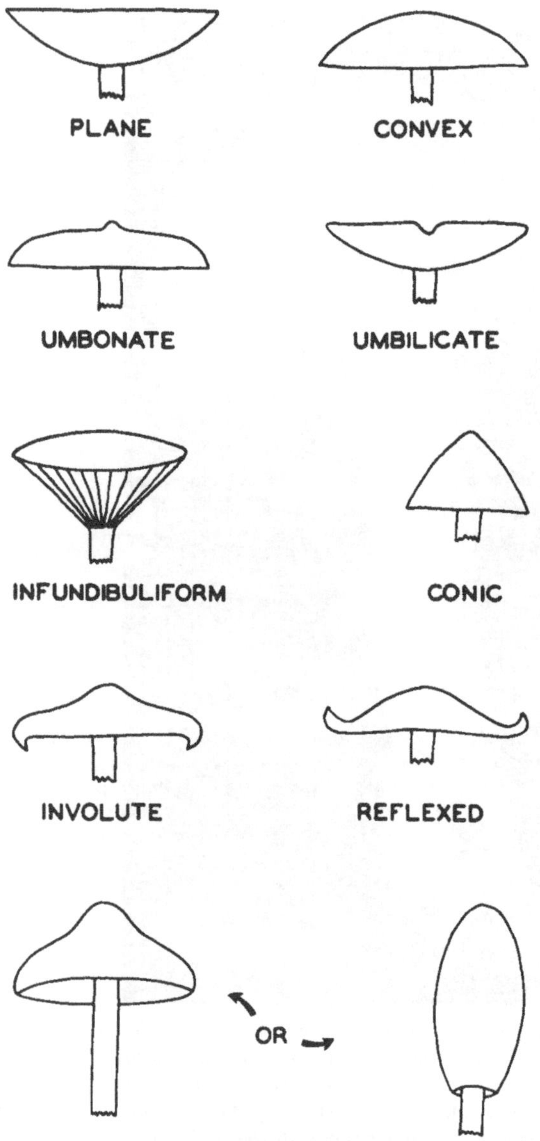

FIGURE 3. Pileus shape is an important characteristic in the identification of gill fungi. Different shapes are diagrammed above.

FIGURE 4. The Shaggy Mane, *Coprinus comatus.* About 90% of natural size.

The characteristics of the stipe are also important in identification. In most agarics the stipe is centrally placed but in some it may be laterally or eccentrically placed (Figure 5). This is also true among the pore fungi; thus, *Boletus edulis* (Figure 35) has a centrally placed stipe, while *Boletinus porosus* (Figure 37) has an eccentric or lateral one. Color, length, thickness, whether hollow or solid, whether or not it is easily separated from the pileus, etc. are all important stipe characteristics that may be used in making an identification.

In the agarics, gill characteristics are also quite important. Number of gills per centimeter (counted at the margin of the pileus) shows quite wide variation between species (Figure 6). Width and color of gills and whether or not they fork are also useful diagnostics. Method of gill attachment is especially important, and the six recognized methods of attachment are diagrammed in Figure 7. Free, ventricose, and adnexed gills are not attached to the stipe, although adnexed gills may grow so close to the stipe that upon superficial examination, they may appear to be attached. Adnate, decurrent, and sinuate gills are attached to the stipe, with decurrent gills extending down the stipe, and sinuate gills being notched, as shown in Figure 7.

The first step that must be made in the identification of any species of gill fungus is that of determining its spore color, and this means the color of a great mass of spores to the unaided eye, not the color of an individual spore as it might be seen under a microscope. Sufficient quantities of spores to enable a proper judgment regarding color are obtained by making a spore print. This is accomplished by removing the stipe from a fresh mushroom, placing the pileus gill side down on a piece of paper, and then covering it with a cup, glass, or beaker to prevent too rapid drying out (Figure 8). Usually within an hour or two (or less if the mushroom is fully mature), enough spores mature and fall to the paper, where they form a dusty deposit, the color of which may be easily determined on sight. It is often recommended that the pileus be placed half on white paper and half on black paper as shown in Figure 8; in this manner, if the mushroom is white-spored, the deposit will be quite obvious on the black paper, and if it is black-spored, the deposit will be quite obvious on the white paper. In Figure 9, spore prints are shown of a white-spored mushroom (*Mycena haemopta*), a purple-brown spored mushroom (*Psathyra umbonata*), and a black-spored mushroom (*Coprinus micaceus*); only the purple-brown spores are seen on both black and white paper. For general use, I prefer to use gray paper for making spore prints. Thus, in Figure 10 may be seen the spore prints of 14 different species of mushrooms representative of the 5 major spore colors, all made on the same piece of gray paper and all easily judged as regards their colors. As far as I am aware, the use of gray paper for making spore prints could lead to only one error, and this is not a serious one. The Oyster Mushroom (*Pleurotus ostreatus*) has a close relative, *Pleurotus sapidus,* which has lilac-colored spores (although it is classified with the white-spored group). A light spore deposit from *P. sapidus* on gray paper might possibly be interpreted as white, but since this

FIGURE 5. Stiped mushrooms may have a centrally placed stipe as those on the left or laterally placed as those on the right.

FIGURE 6. Method of measuring the number of gills per centimeter at the margin of the pileus. In pileus *A* there are 11 gills per centimeter; in *B*, there are 8.

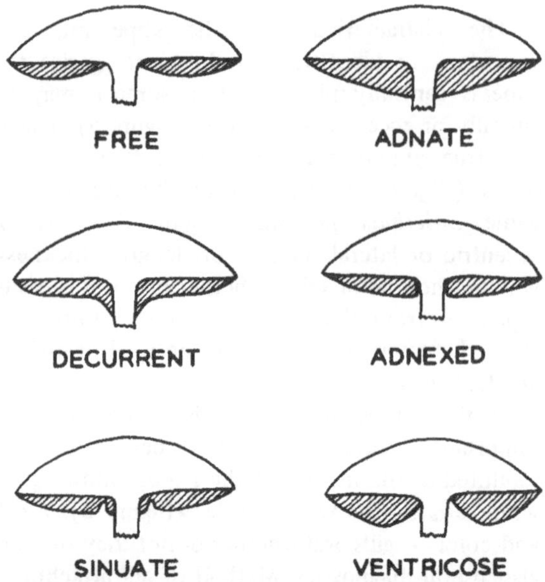

FIGURE 7. Methods of gill attachments in agarics.

FIGURE 8. Method of preparing a spore print. It is often recommended that both black and white paper be used for catching the spore deposit. In the above figure a pileus of *Pholiota adiposa* has been used.

would lead the collector to identify it as *P. ostreatus*, no harm would be done since both are edible. In fact, these species are otherwise so similar that there is some question as to whether or not they should actually be classified as different species.

The great majority of gill fungi have spores that are easily classed as having one of five colors if proper spore prints are made. These colors are white, black, purple-brown, yellow-brown, and pink. There are a few exceptional species which have colors different from these five, but such species are few in number. The lilac-spored *Pleurotus sapidus* mentioned above is such an exception. The sometimes poisonous *Lepiota morgani* has green spores and thus is easily

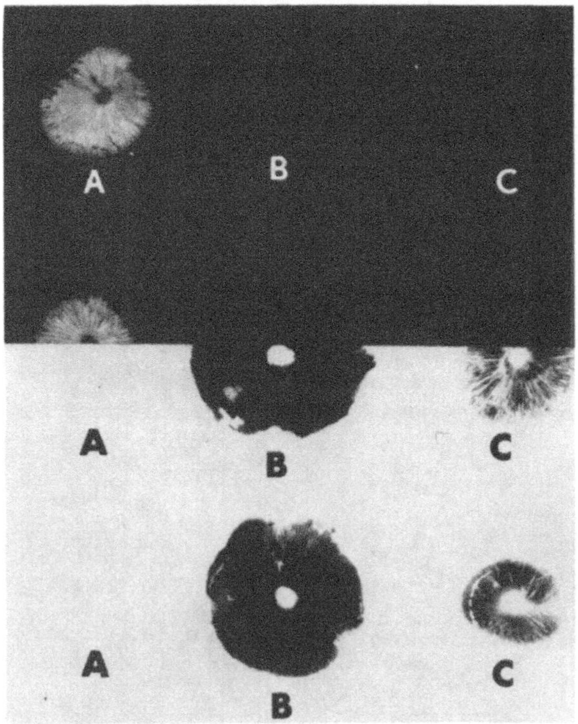

FIGURE 9. Spore prints on both black and white paper. Prints marked *A* are from a white-spored mushroom – *Mycena haemopta* – visible on black but not on white paper. Prints marked *B* are from *Psathyra umbonata*, a species with purple-brown spores – visible on either black or white paper. Prints marked *C* are from a black-spored mushroom – *Coprinus micaceus* – visible on white but not on black paper.

identified, but only through a key if green is considered to be white (shades of George Orwell!). Some species of *Russula* have yellow spores, but these, like *L. morgani*, are classified with the white-spored agarics.

Folklore abounds with easy "surefire" methods of distinguishing poisonous mushrooms from edible ones; such methods should be discarded at the outset since none of them are reliable. For example, it has often been stated that if poisonous mushrooms are cooked with a silver spoon, the spoon will blacken, while edible mushrooms will not have such an effect; sole and repeated reliance on this method is a fairly certain way of getting poisoned! Another superstition is that poisonous mushrooms do not grow on wood and do not grow in clumps. While it is true that most poisonous mushrooms do not grow in clumps or on wood, there is certainly one notable exception: the Jack-o-Lantern (*Clitocybe illudens*) shown in Figure 64 grows in clumps on wood and is poisonous. The upper surface layer (cuticle) can be easily peeled from some mushrooms, while from others it cannot. Whether or not the cuticle can be peeled from the pileus is a specific characteristic and not one that is in any way related to whether or not the mushroom is edible or poisonous. There is a common belief that any mushroom with a volva is poisonous and some are; however, the edible wild Silky Volvaria (Figure 17) has a volva, as does the Padi Straw Mushroom which is grown commercially in the Orient and eaten by millions of people. For a more thorough discussion of the fallacy of believing that these legends provide reliable methods for distinguishing between edible and poisonous mushrooms, the work of Herrick[95] is recommended.

The only reliable method for identifying an unknown mushroom is through proper use of a reliable key followed by a comparison of the characteristics of the specimen with a written description of the species and also a comparison of the specimen with a good illustration. The procedure might well be as follows: let us assume that in late July the collector finds a type of mushroom growing singly or scattered on a decaying hardwood log and that his only manual is Christensen's[37] *Common Fleshy Fungi*. Most of the pilei are convex (Figure 11), although some of the larger, more mature ones may be almost plane and some of the younger ones may appear almost campanulate. The stipe is removed from one mushroom, the pileus placed gills down, half on black paper and half on white paper (or preferably gray paper) and then covered with an ordinary drinking glass, cup, or bowl. Within a few hours or less enough spores have fallen so that it can be easily seen that, *en masse*, the spores are pink in color (Figure 12).

The collector then finds that early in his manual (page 14) there is a key to the genera of pink-spored mushrooms. The first two choices offered him in the key are

1. Stem lateral or absent, on wood . *Claudopus*

OR

1. Stem central 2

Obviously the stipe is centrally placed (Figure 11) and so his choice here is an easy one and he proceeds as directed in the key to the second set of statements:

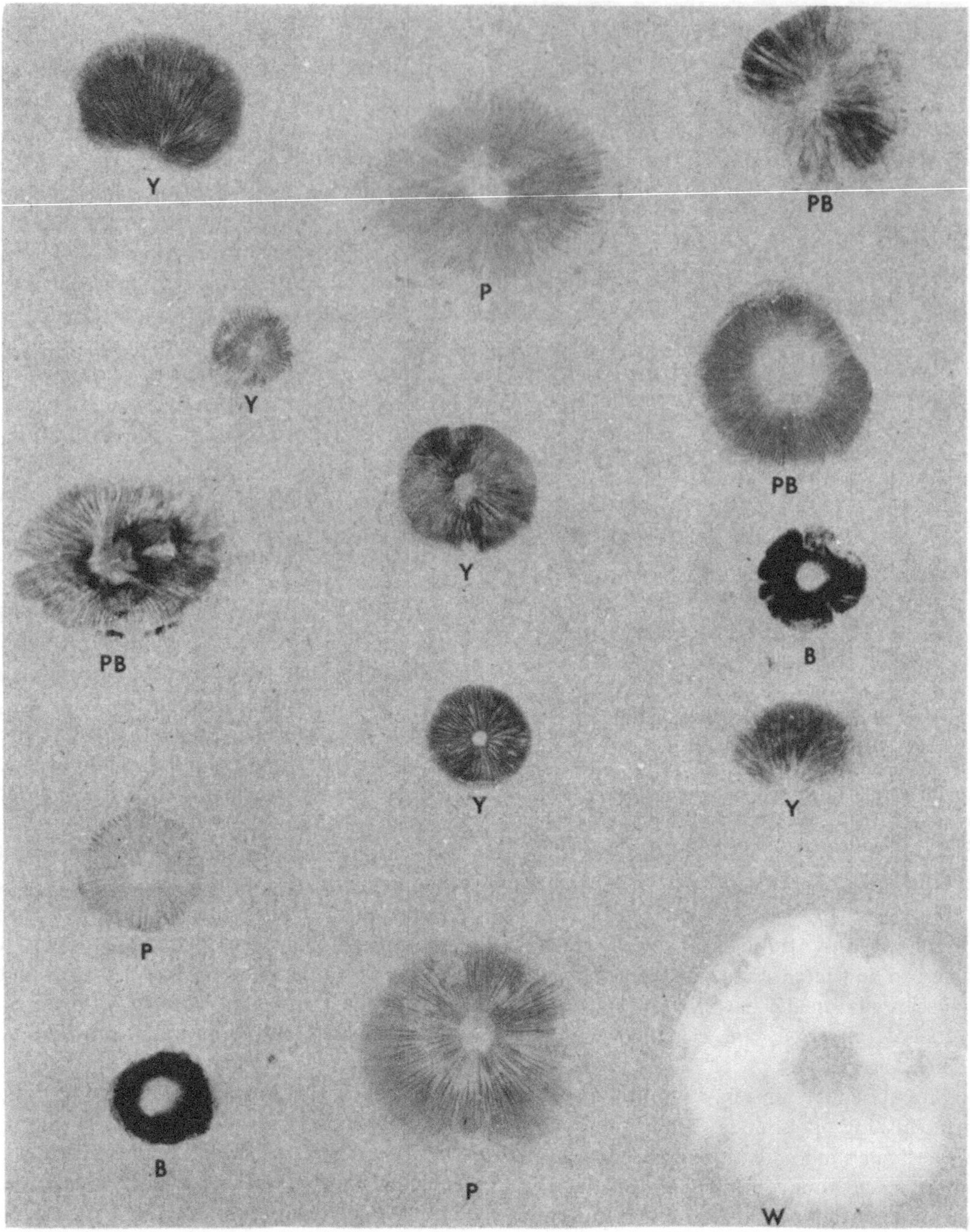

FIGURE 10. Spore prints on gray paper. The first step in identifying a gill fungus is to determine the color of the spores en masse. In general, light gray paper is best for making spore prints. In the above figure all of the five basic spore colors are visible on gray paper: black – B, white – W, pink – P, yellow-brown – Y, purple-brown – PB.

FIGURE 11. The Fawn-Colored Mushroom, *Pluteus cervinus.* (Photograph by G. F. Atkinson.)

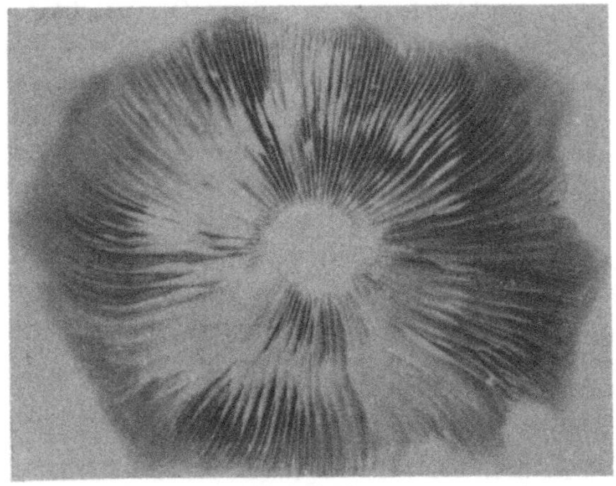

FIGURE 12. Spore print of *Pluteus cervinus. En masse* the spores of this species are pink.

2. Cup at the base of the stem, no ring on the stem. . . . . *Volvaria*

OR

2. Neither cup nor ring present 3

Again the choice is an easy one and there is no room for confusion. Examination of his specimens shows no trace of either an annulus on the stipe or a volva at the base, so he proceeds as directed to the third set of statements:

3. Gills free, cap separating easily from the stem . . . . *Pluteus*

OR

3. Gills not free. 4

He finds that the stipe can be rather easily separated from the pileus and that when he splits a specimen longitudinally through the center as shown in Figure 13, the gills are not attached to the stipe. At this point it would appear that he has found a species of the genus *Pluteus*. He may then turn to page 119, where he finds a brief description of the genus and a key to the species of *Pluteus*. The generic description fits his specimen and so he turns next to the specific key which lists four species:

1. Cap pale tan to brown, silky shining, rarely faintly scaly, common . . . . *P. cervinus*

OR

1. Cap white, covered with fine hairs or tufts of hair . . . . *P. tomentosulus*

OR

1. Cap brown with radial ridges in the center. . . . . *P. nanus*

OR

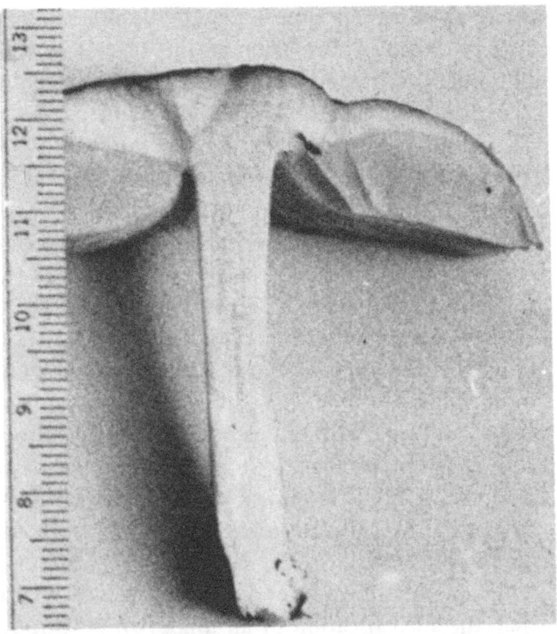

FIGURE 13. Longitudinal section through basidiocarp of *Pluteus cervinus* showing method of measuring gill width.

1. Cap bright lemon yellow . . . . . . *P. leoninus*

On the basis of the descriptions of the pilei, it is obvious that the collector has probably found *Pluteus cervinus*, the Fawn-Colored Mushroom. However, he should not assume this until he has read the description of this species and satisfied himself that the species description fits the specimens that he has found. Thus, he should determine if, as Christensen states, the pileus is 5 to 14 cm wide and is campanulate, convex, or plane. He should also note if the cuticle can be easily peeled from the pileus (Figure 14), a characteristic of this species although it is not unique in this respect. He should measure the width of the gills (Figure 13) and the thickness of the flesh above the gills; he should also determine that the number of gills is 10 to 15/cm at the margin, that in young specimens they are white in color, but that in older specimens thay may be flesh color or tan. He should note whether the stipes are in the range of 6 to 15 cm in length, white toward the top and tan below, and the outer rind of the stipe is brittle-tough with a spongy interior. The thickness of the stipe of this species varies over a rather wide range, 0.6 to 2.5 cm. The specimens shown in Figure 11 have rather narrow stipes but, not uncommonly, stipes up to an inch in diameter may be found. When first encountered the thick-stiped specimens may be thought to be a different species, but examination reveals that in all other characteristics except stipe diameter specimens are very similar. If the specimens fit the description in all respects, the collector can be sure that he has found *Pluteus cervinus*. He will note that his manual lists this as an edible species.

Pluteus cervinus, the Fawn-Colored Mushroom, is an easily identifiable species and one that the amateur should quickly learn to recognize. The pileus is edible and quite tasty, but because of its toughness the stipe should be discarded. It may be found until October and once learned, cannot be easily confused with any other species. Because it will grow on sawdust, the suggestion has been made that this is a species which might easily be brought into cultivation on this waste from the lumber industry.

From the above brief account it would appear that the identification of an unknown mushroom is a straightforward and relatively easy process, and so it is in many instances. However, the amateur (or the professional, for that matter) will encounter specimens which are not quite so easily identified as was *Pluteus cervinus* in the example described above. This may be due to a variation in the specimen collected (it is always best to collect a number of specimens), to the fact that two species may vary one from the other only with respect to minor differences, or it may well be due to the fact that while each statement in a key may be perfectly clear to the person who wrote it, it may not be quite as clear to the reader with little experience in identification work. Repeated use of the same key will usually clear up such confusion

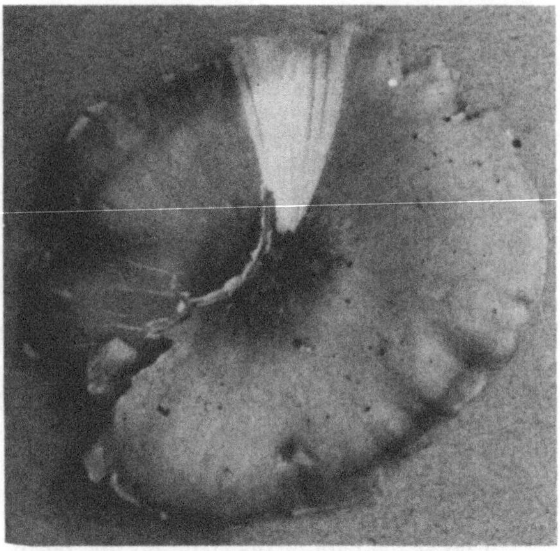

FIGURE 14. Pileus of *Pluteus cervinus* showing that cuticle may be easily peeled from the surface.

and with repeated experiences with the key the user will soon come to know exactly what the writer had in mind when he wrote each statement.

Use of several keys will soon show the user that even professional taxonomic mycologists do not always agree as to what the correct scientific name of a particular species is. Indeed, Wasson[190] has suggested that perhaps the small band of specialists in the field of mushrooms has been overly preoccupied with problems of taxonomy, classification, and nomenclature. A case in point is provided by the Sulphur Fungus, which is discussed later in this section. This is one of the few species in the family Polyporaceae (pore fungi) which is tender enough to eat and, once seen, is difficult to confuse with any other fungus. For many years it was named *Polyporus sulphureus* but now may also be found under the name *Laetiporus sulphureus*. The name change was probably made for sound taxonomic reasons, but in most instances such technical points as these need not concern the amateur. To him the important thing is to be able to recognize this particular fungus, and if he can, it is of little importance to him whether it is called *Polyporus, Laetiporus*, or, for that matter, Charlie. Presumably, all of these taxonomic problems and questions of nomenclature will eventually be solved, but the amateur mycologist or mycophagist need not wait for their solution to accomplish what he has in mind.

No attempt to key out additional species will

be made here. On the contrary, a number of common, rather characteristic edible wild mushroom species will be figured and described as an aid to the amateur mycologist. The species listed are those which can usually be identified without resorting to the use of microscopic characteristics. It should be emphasized that these are aids only and that before eating any mushroom hitherto unknown to him, the collector should identify it by use of a reliable key and should then compare his specimen with the description given in the manual. Wherever possible, poisonous or inedible species that might possibly be confused with the species under consideration will be illustrated and described, and specific points to look for in order to avoid misidentifications will be noted. By far the largest number of species discussed in this connection are gill fungi, the true agarics, since there are far more edible species in this group than there are in other groups of fungi. However, a few such other fungi will be discussed, notably the Morels, a few pore fungi, and certain of the true puffballs.

The Rooting Collybia (*Collybia radicata*)

Most mushrooms have a stipe that ends at ground level or extends only slightly below ground level, beyond which point the only fungal tissue that is evident consists of rhizomorphs (closely compacted bundles of hyphae) or quantities of many-branched, usually white mycelium. The Rooting Collybia is an exception to this general rule and hence this species, *Collybia radicata*, is easily separated from all other species of mushrooms. The stipe of *C. radicata* continues far underground, tapering gradually to form a structure that superficially resembles the tap root of many flowering plants (Figure 15). Thus if a mushroom belonging to this species is carefully removed from the soil, not uncommonly as much of the stipe may be found below the soil as above the soil.

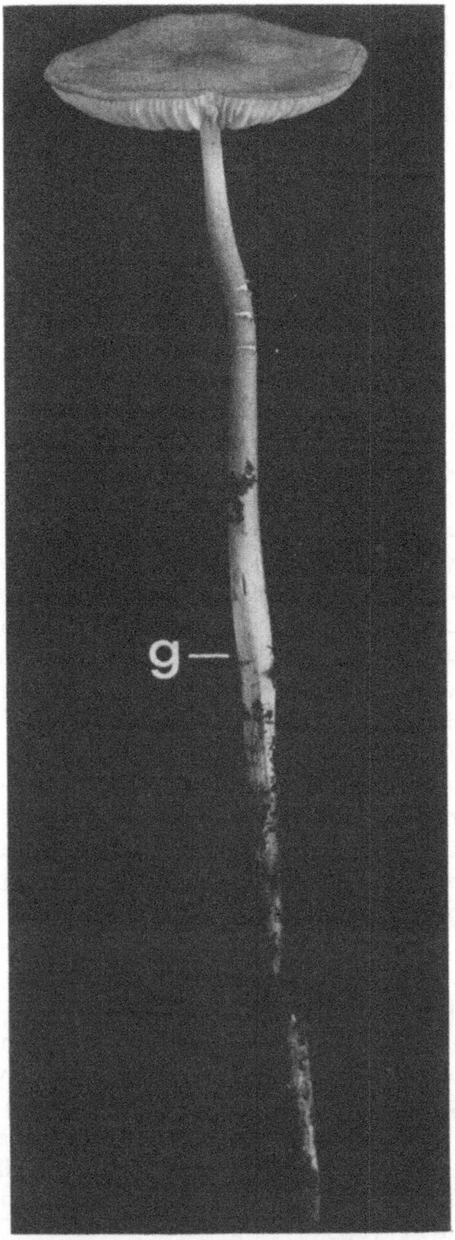

FIGURE 15. The Rooted Collybia, *Collybia radicata.* Position *g* indicates approximate ground level. (Photograph by G. F. Atkinson.)

The *pileus* varies in size from 1 to 4 in. in diameter, is thin, and convex or nearly plane. In some specimens it may be umbonate. It is grayish, grayish-brown, or smoky in color. The surface of the pileus is generally radiately wrinkled or furrowed and may be dry or very sticky. In fresh specimens the pileus is usually sticky but in older specimens or during a very dry season the pileus is quite dry.

The *gills* are not closely packed – about 10/cm at the margin of the pileus. They are 5 to 15 mm wide and are white to pale yellow in color. Gills are adnexed (see Figure 7) or short decurrent. The flesh above the gills is thin and white.

The *spores* of this species are white, just as in the deadly Amanitas.

The *stipe* is 2 to 8 in. long above ground and may sometimes extend an equal distance below ground. In my experience, the stipe is usually 6 to

10 mm thick; however, Christensen[37] reports stipe thicknesses of 3 to 16 mm.

The Rooting Collybia may be found occurring singly or in small groups (but not in clumps), often near hardwood stumps or trees, found from June to October. Although all characteristics of a specimen of this mushroom must be determined, the principal distinguishing feature is the long, subterranean, taproot-like extension of the stipe.

The Velvet-stemmed Mushroom [*Collybia (Flammulina) velutipes*]

The dedicated mycophagist should by all means become acquainted with this species because, unless weather conditions are too severe, it may be found during any month of the year. The Velvet-Stemmed Mushroom, *Collybia velutipes* (also called *Flammulina velutipes*), is related to the species just described (*C. radicata*), but it would be impossible to confuse these two species because they are different in color, growth habit, size, and habitats.

The *pileus* of *C. velutipes* is 2 to 6 cm in diameter and hence this species is generally smaller than *C. radicata*. While it may be plane, in most specimens it is convex with the margin curved down. It is tan to reddish-brown in color, with the color usually being paler at the margin. In fresh specimens the pileus is sticky and appears moist and shiny.

The *gills* are more closely spaced than in *C. radicata*, being 20 to 25/cm at the margin and adnexed or toothed at the margin. They are white or pale tan in color and if examined with a lens, the edges are seen to be fringed. The flesh above the gills is thin and is the same color as the gills.

The *spores* are white, as in all species of the genus *Collybia*.

The *stipe* is 3 to 10 mm thick and 3 to 10 cm long. It is commonly curved unless growing from a horizontal surface of a stump or log. It is pale yellow in very young specimens but the upper part later becomes tan and the lower part becomes dark in color due to the presence of dense, velvety, reddish-brown or purplish hairs. It is this latter characteristic which makes this species easily recognizable and, of course, gave rise to its common name. There is no annulus or volva.

Unlike *C. radicata,* the present species does not grow on the soil but is found on decaying logs and stumps and occasionally on living deciduous trees. Also, unlike *C. radicata, C. velutipes* does not occur singly; large numbers of stipes arise close together and so it is to be found growing in clumps (Figure 16), a characteristic which is of obvious advantage to the individual who is collecting for the table. The three main distinguishing features are (1) the dark, velvety appearance of the lower portions of the stipes, (2) the tan to reddish-brown, usually sticky pilei, and (3) the habit of growing on decaying wood in clumps. The poisonous Jack-o-Lantern, *Clitocybe illudens* (Figure 64), also grows in clumps on stumps and logs, but the two species differ in size and color, and the gills of *C. illudens* are definitely decurrent. For a more complete discussion of *C. illudens,* see the discussion of *Armillaria mellea* (Honey Mushroom), another wood-inhabiting species, and the section on poisonous mushrooms.

There is a distinct advantage to the mycophagist in being able to recognize *C. velutipes* since, although it is not a choice mushroom, it may be found at times of the year when other mushrooms cannot be found. During the warmer months of the year, this species like other species of mushrooms may be found to be infested with small insects and slugs, but during the winter months it is completely free of such small animal life. Some individuals find the sticky cap distasteful, in which case the cuticle should be removed from the pileus. Certain Polish-Americans of northern Illinois especially prize the Velvet-Stemmed Mushroom and refer to it as "veshpoofki" (Blei[21]).

The Silky Volvaria (*Volvaria bombycina*)

The Silky Volvaria is an edible species, but here it is recommended for eating only on the assumption that every diagnostic characteristic will be carefully checked. The generic name, *Volvaria,* would imply that a volva is present, and since volvas are common to all species of *Amanita* (Figures 2, 58, 59, 60, 61, and 63), many of which are poisonous, it might seem the better part of good judgment not to recommend any species of *Volvaria* for eating. However, the Silky Volvaria is quite tasty, and the two genera are quite distinct, so if reasonable care is exercised there should be no problem of confusing one with the other.

Both Krieger[110] and Christensen[37] report that the *pileus* may be up to 8 in. in diameter; however, I have never seen one larger than 4 in. In young

FIGURE 16. Clump of basidiocarps of the Velvet-Stemmed Mushroom, *Collybia velutipes*. Note that the gills are free and that the basal portions of the stipes have a dark, velvety appearance. (Photograph by G. F. Atkinson.)

specimens the pileus is conical, but as it expands it becomes more broadly conical or convex. The fresh pileus is snow-white in color (Figure 17) and is densely covered with small tufts of hairs or scales, all pointing downward and giving the pileus a distinctive silky appearance.

The *gills* are 8 to 12 mm wide, are free, and number 20 to 25 at the margin. They are white at first but later (as the spores mature) they become pink or flesh-colored. The flesh above the gills is thin and white.

The *spores* are pink. This is a very important point, since the spores of the poisonous Amanitas are all white.

The *stipe* is 2 to 6 in. long, varying in thickness from 1 to 1.5 cm at the top. It is tan at the base, white at the top, solid, and tapers upward. It is not bulbous at the base. There is no *annulus*, another important difference between *Volvaria* and *Amanita*.

The *volva* is very large and conspicuous (Figure 17) and in all specimens I have seen, could never be confused with the sometimes inconspicuous volva not infrequently seen in *Amanita* (Figure 2). The volva is tough and leathery and white to tan in color.

Christensen[37] states that *V. bombycina* may be solitary or scattered in groups of up to a dozen basidiocarps on living trees or hardwood logs, principally box elder, maple, or beech. However, Krieger[110] states that it is usually solitary and grows from wounds in the bark of a variety of deciduous trees, but commonly on maple. I have found it only on logs and solitary.

The principal distinguishing features of the Silky Volvaria are (1) the snow-white pileus with a silky sheen, (2) the large and very obvious volva, (3) the complete absence of an annulus on the stipe, and (4) the definitely pink spores. The last feature is an extremely important characteristic, since the Amanitas which are also volvate have white spores.

Kauffman[106] reported that two species of *Volvaria* are poisonous: *V. speciosa* and *V. gloicephala*. However, it would be very difficult to confuse either of these species with *V. bombycina*, since both are smaller than *V. bombycina* and the pilei are described as glabrous (smooth, without hairs) and viscid. He noted that the pileus of *V. speciosa* is white or tinged gray and that the species occurs infrequently, and the pileus of *V. gloiocephala* is smoky-gray to pearl-gray in color and that the species is rare.

FIGURE 17. The Silky Volvaria, *Volvaria bombycina*, natural size. (Photograph by G. F. Atkinson.)

The Meadow or Field Mushroom (*Agaricus campestris*)

The common Meadow or Field Mushroom is a species that has been eaten for many years. For a period of time, after having been called *Agaricus campestris* for many years, the name was changed to *Psalliota campestris*, but now it seems to be again generally referred to as *A. campestris*, except, perhaps, in Europe. It was at first thought that the Common Cultivated Mushroom was this species, but this latter is now generally considered to be a distinct species (*A. bispora*) or a variety of the Meadow Mushroom, *A. campestris* var. *bispora*. As far as I have been able to determine, *A. campestris* has never been brought into cultivation, at least not on a commercial scale.

The *pileus* of the Meadow Mushroom varies in size from 1 1/4 to 4 in. in diameter and varies both in color and texture of the surface. It may be white to tan in color and the surface may be smooth, silky, or covered with small flat scales. It is usually convex when young but upon expanding may become almost plane. As the young pileus expands, the loose tissue covering the gills breaks in such manner that a *veil* is left hanging around the margin of the pileus (Figure 18). The cuticle usually may be peeled off easily, a characteristic of this well-known species that may have contributed to the development of the myth that edible mushrooms are easily peeled, while poisonous ones are not.

Gills apparently are variable, Christensen[37] describing them as free, and Krieger[110] describing them as ventricose. They are 6 to 12 mm in width and are 15 to 20/cm in number at the margin of the pileus. They are white at first, then turn pink and finally appear purple-brown when the spores mature. The flesh above the gills is firm, white in color, and may be 1 to 2 cm thick in the stipe area.

The *spores* of *A. campestris* are purple-brown in color.

The *stipe* is 2 to 4 in. long and is 1 to 2 cm thick, sometimes slightly thicker at the base. It is smooth or nearly smooth and white or whitish in color. In most specimens, the *annulus* is at about the middle of the stipe, is white, thick, and quite obvious. There is no volva. Thus, while the poisonous Amanitas have both volva and annulus, *Volvaria* has only a volva, and *A. campestris* has only an annulus.

The Meadow Mushroom may sometimes occur in "fairy rings" and often is to be found in pastures, manured ground, or on golf courses. According to Krieger,[110] it never occurs in woods. It may occur scattered or in groups but I have never seen it in clumps. It may be found in late summer and early autumn. It is an exceptionally fine flavored mushroom, especially when fully matured and the gills have begun to darken.

The principal distinguishing features of the Meadow or Field Mushroom are (1) the absence of a volva, (2) a usually distinct annulus, (3) a veil hanging from the margin of the pileus, and (4) purple-brown spores. It is especially important to determine spore color, although the mere fact that a mushroom has purple-brown spores does not mean that it may not be poisonous. Thus, Smith[171] listed *Naematoloma fasciculare*, Atkinson[9] listed *Stropharia coronilla*, and Lange and Hora[112] listed *Stropharia hornemanii* as

FIGURE 18. The Common Field Mushroom, *Agaricus campestris.* Note the prominent annulus and the veil hanging from the margin of the pileus. (Photograph by G. F. Atkinson.)

poisonous, all three species being purple-brown spored. There should be no reason for confusing these species, however, since the gills of no species of *Stropharia* are free and the pilei are usually viscid. The gills of *N. fasciculare* are also not free, and a true annulus is lacking in this species.

The Smooth Lepiota (*Lepiota naucina*)

The Smooth Lepiota, *Lepiota naucina,* is recommended here for eating with some reservations, not because there has been any suggestion that it is poisonous, since it is, in fact, a rather tasty species and is often found in considerable abundance in late summer and early autumn. It is recommended with reservations because in a cursory examination, it would be possible to mistake one of the poisonous white Amanitas (Figures 58, 59, and 60) for this species. Nonetheless, if the species description presented here is applied to every single specimen prepared for eating, there should be no difficulties.

The *pileus* of *L. naucina* is oval or globose before expansion and becomes broadly convex to plane at maturity. The expanded pilei may range in size from 2 to 4 in.; the unexpanded ones are somewhat smaller. The color is white or nearly white, and sometimes very pale tan. The surface is quite smooth and while it may sometimes be irregularly cracked, it is never scaly (Figure 19).

The *gills* of this species are free and white when young. They then become pinkish and finally light brown. They are 5 to 10 mm wide and about 15/cm in number at the margin. The flesh above the gills is firm in texture and white.

FIGURE 19. Basidiocarps of the Smooth Lepiota, *Lepiota naucina.* Both stipe and pileus are white and smooth. Note bulbous enlargement at base of stipe and the well-developed annulus. A characteristic of the species is the ease with which the stipe can be separated from the pileus.

The *spores* of *L. naucina* are white.

The *stipe* is 2 to 4 in. long and 6 to 12 mm thick at the top. It is white in color, hollow when mature, and gradually tapers upward from a bulbous base. The bulbous part of the base is quite smooth (Figures 19 and 20) as is the rest of the stipe, and there is no evidence of a volva, even one reduced to scales, as is often seen in various Amanitas. A characteristic, although not unique, feature of this species is the mode of attachment of the stipe to the pileus and the ease with which it can be detached from the pileus. If the pileus of a mature specimen is held firmly in one hand and the basal part of the stipe pushed to one side, it usually parts from the pileus immediately, leaving a ball-like portion on the top of the stipe and a corresponding socket-like opening in the lower part of the pileus (Figure 20). This species has a prominent *annulus* (as do all Lepiotas), usually placed above the middle of the stipe. It is white in color, firm, and sticks straight out from the stipe. As in other members of the genus *Lepiota*, and especially in mature specimens, the annulus is so loose that it may be moved up and down (Figure 20).

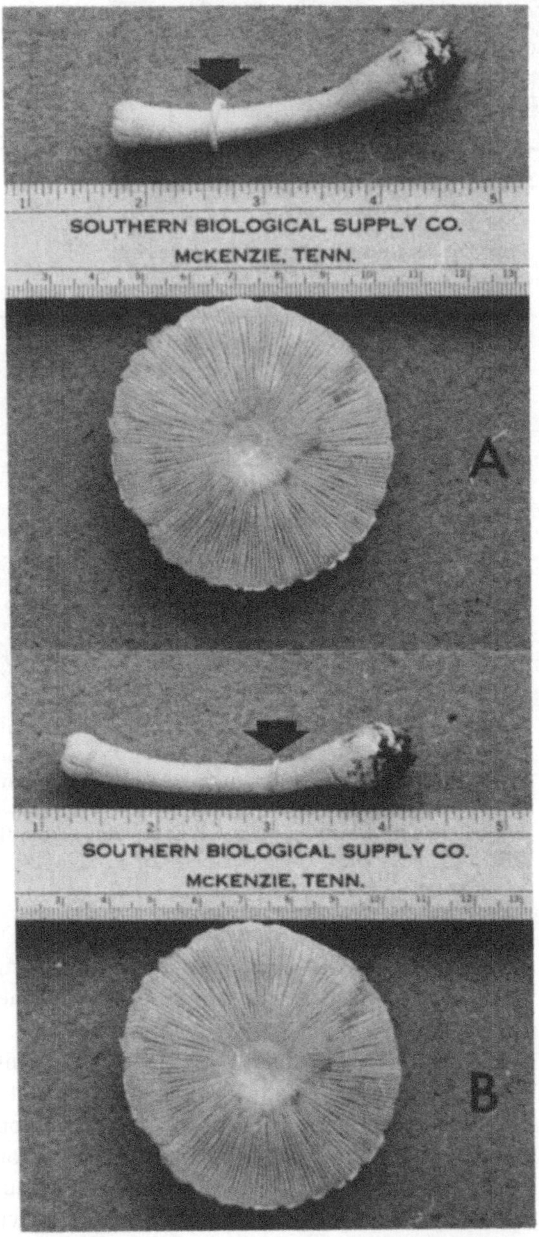

FIGURE 20. Among the distinguishing features of *Lepiota naucina* are the movable annulus and the ease with which the stipe separates from the pileus. In *A* the annulus is shown in the position in which it developed; in *B* it has been moved to a position farther down the stipe. Stipe-pileus connection resembles a ball and socket joint.

The Smooth Lepiota grows on the ground in fields, pastures, and open woods. The author has found this species most abundant in grassy areas at and just inside the edge of open woods, where as many as 40 or more mushrooms may be found in a relatively small area. When such a site is found, it is worthwhile to revisit it at intervals of several days since several crops are usually produced. It is sometimes found in abundance along roadsides, but if it is collected from such habitats the collector should satisfy himself that herbicidal sprays have not been used in the area.

The principal distinguishing features of *L. naucina* (in addition to its having white spores) are (1) it is white and the surfaces of both pileus and stipe are smooth, (2) the annulus is prominent, sticks straight out from the stipe, and, at least in more mature specimens, may be moved up and down easily, (3) the base of the stipe is bulbous and is smooth without a trace of scales or other tissue that might indicate the presence of an inconspicuous volva, and (4) the stipe parts easily from the pileus to which it is attached by what appears to be a ball and socket arrangement. When collecting the Smooth Lepiota for the table, the above four criteria should be applied individually to each specimen, since it sometimes happens that a mushroom or two of one of the poisonous white Amanitas may be found in the same general habitat as this edible species. Therefore, as each specimen is found, the entire mushroom must be collected so that the smooth, bulbous, basal portion of each stipe may be examined. The ease of separation of stipe and pileus should be tested when the collector has satisfied himself about the non-volvate nature of the smooth, bulbous base. The stipes may then be discarded since they are hollow and somewhat tough when mature.

The Parasol Mushroom (*Lepiota procera*)

Although Krieger[110] described the Parasol Mushroom (Figure 21) as one of the finest of all edible mushrooms, and others have also attested to its tastiness, as with its close relative, the Smooth Lepiota, it is recommended here with some reservations, and extreme caution is advised, not because there is any suggestion that it may be poisonous, but rather because it seems possible for the beginner to confuse poisonous species with this edible one. For example, Krieger notes that it

FIGURE 21. The Parasol Mushroom, *Lepiota procera*. (Photograph by G. F. Atkinson.)

may be confused with the poisonous *Lepiota morgani,* although this misidentification could occur only if no spore print were made, since Morgan's Lepiota has green spores.

The *pileus* varies in diameter from about 2 to 6 in. Its shape has been variously described as oval or ovate, campanulate, or convex to flat with an umbo. The specimen shown in Figure 21 would best be described as convex, although there is a slight indication of its being broadly umbonate. At first the pileus is uniformly reddish-brown in color, but upon expansion, the surface breaks up into concentrically arranged brown scales which may be interspersed with white, woolly scales.

The *gills* are free, about 6 to 12 mm wide at their widest point which is near the middle, and number about 15/cm at the margin. They are white or pinkish when young, becoming pale brown with maturity. The flesh above the gills is white and thick.

The *spores* of *L. procera* are white.

The *stipe* is typically long and relatively narrow, ranging in length from 6 to 10 in. or possibly longer and about 1/4 to 1/2 in. thick. As in *L. naucina,* the base of the stipe is bulbous and the stipe gradually tapers upward from it. The *annulus* is thick, quite obvious, and is placed relatively high on the stipe; it is white on the upper side but covered with brown scales on the lower side. The annulus of the Parasol Mushroom is movable just as it is in both the Smooth Lepiota and Morgan's Lepiota. The stipe sets in a socket in the lower part of the pileus.

The Parasol Mushroom may be found on the ground, singly or scattered, during late summer and early autumn in pastures, along roadsides, or in open deciduous woods. Again, this mushroom species is not recommended for the beginner. On the contrary, no collector should eat what he believes to be this species until he is quite sure that he is familiar with *L. morgani* as well as the various poisonous Amanitas. The spore color should be determined and the bulbous bases of the stipes carefully scrutinized in order to determine if there are any traces of what might be a volva.

The Oyster Mushroom (*Pleurotus ostreatus*)

The Oyster Mushroom, *Pleurotus ostreatus* (Figure 22), is a white-spored species which is easily recognized and in no way resembles the deadly white-spored Amanitas. There are several species which it superficially resembles, but learning to distinguish the Oyster Mushroom from others may be quickly accomplished simply by seeing it a few times.

Both Christensen[37] and Krieger[110] list *pileus* diameters for this species as ranging from 2 to 5 in ; however, under ideal conditions of moisture and temperature the pilei may be as large as 6 or 7 in. When collecting for the table, pilei in the range of 2 to 3 in. are preferable because of the greater ease with which they may be prepared for cooking. The surface of the pileus is smooth, moist appearing, and is usually whitish or grayish in color although it may sometimes be lightly tinged with brown.

The *gills* are decurrent (extending down the stipe), are white in color, and are 5 to 15 mm wide. They are about 18 to 20/cm in number at the margin. The flesh above the gills is white.

The *spores* of the Oyster Mushroom are white. A closely allied species, *Pleurotus sapidus,* has lilac-colored spores, but if this latter species is confused with *P. ostreatus,* no harm will be done

FIGURE 22. Cluster of small basidiocarps in the Oyster Mushroom, *Pleurotus ostreatus.* This mushroom is white-spored, has short eccentric stipes, and may occur in clusters that weigh several pounds. (From Gray.[79] By permission of Holt, Rinehart and Winston.)

since it too is edible and its taste is quite similar to that of the Oyster Mushroom.

The *stipe* is lateral or eccentric and usually no more than 1 in. long. It may sometimes be hairy at the base. This mushroom commonly grows in clumps, with the stipes close together and attached to a common base, a favorite habitat for small insects. Several other common mushrooms have a superficially similar growth habit, but these (*Trogia, Claudopus,* and *Crepidotus*) will not be confused with *P. ostreatus* once the latter has been seen in typical form. None of the three mentioned attain the size of the Oyster Mushroom, and the spore colors of *Claudopus* and *Crepidotus* are not white; *Claudopus* has pink spores while those of *Crepidotus* are yellow-brown. Like *P. ostreatus, Trogia* has white spores, but the common species of this latter genus (*T. crispa*) has small pilei (1 to 3 cm), is covered with fine hairs, and the margins are lobed or wavy (Figure 23).

The Oyster Mushroom commonly grows on hardwood logs and stumps, and Krieger[110] reported that it may be found on living trees. I have never seen it growing on a sound tree, only on living trees that have suffered previous damage. It is most common in summer and autumn (but only if there is sufficient moisture), but may sometimes be found during the winter months after several days of higher temperatures. This species is often found in great clumps, sometimes large enough to almost fill a bushel basket. There are some differences of opinion regarding the origin of the common name, some claiming that the pileus resembles an oyster, others claiming that this mushroom has the taste of oysters. When the Oyster Mushroom is found in abundance on a large log or stump, it is worthwhile to remember the location and to revisit the site at intervals since several crops may be subsequently produced.

The Fairy Ring Mushroom (*Marasmius oreades*)

The Fairy Ring Mushroom, *Marasmius oreades,* is described by Christensen[37] as one of the best of the edible mushrooms, and Buller[31] considered it desirable enough to attempt to bring it into cultivation. While a variety of mushrooms may often be found growing in circles, apparently *Marasmius oreades* forms these "fairy rings" rather

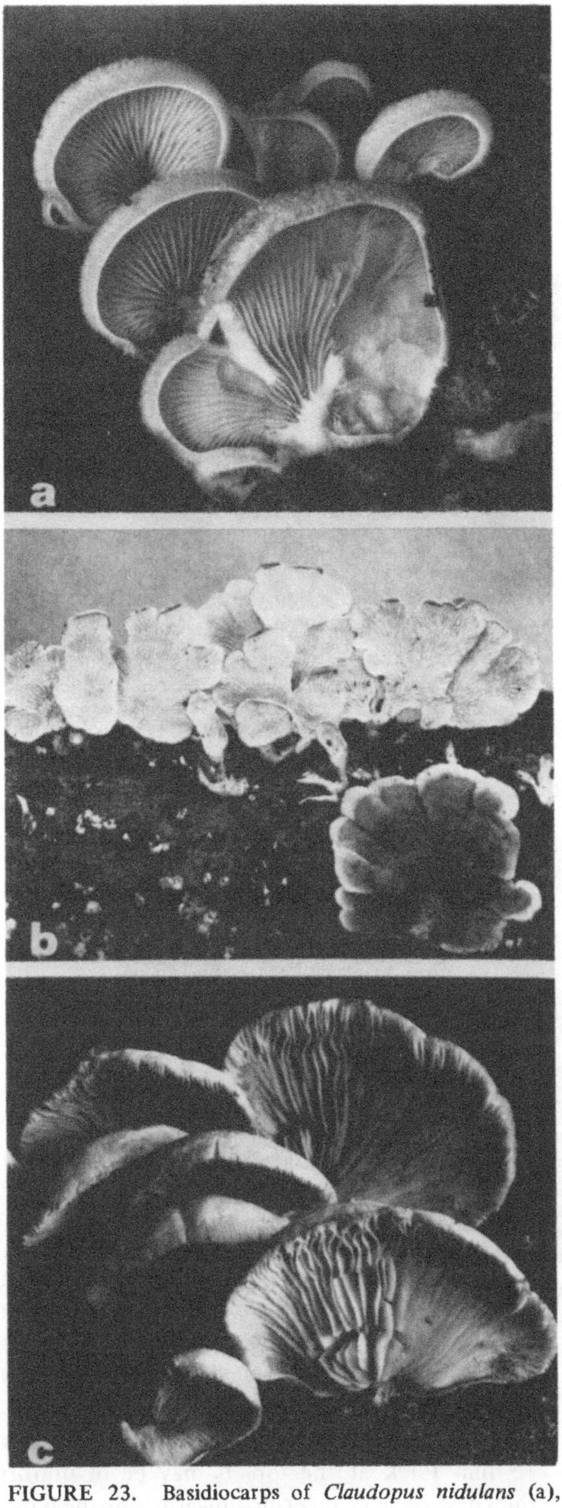

FIGURE 23. Basidiocarps of *Claudopus nidulans* (a), *Trogia crispa* (b), and *Crepidotus malachius* (c). (From Christensen.[37] By permission of Burgess Publishing Company.)

more frequently than most other species and hence derives its common name from that habit.

The *pileus* is usually 1 1/2 to 2 in. in diameter (Figure 24), although smaller ones may be found. It is convex to almost plane and in some instances umbonate; although it is broadly umbonate (see *u* in Figure 24), this characteristic is often not too obvious. Its color varies from tan to nearly white or sometimes pale reddish. The surface of the pileus is smooth.

The *gills* are not spaced close together and at the margin of the pileus short gills may be seen to be rather regularly interspersed with the long gills that originate near the stipe (*g* in Figure 24). The *gills* are free, and the flesh above them is light tan to white; the flesh may be thick near the stipe, especially if the pileus is umbonate.

The *spores* of the Fairy Ring Mushroom are white, but since this species has no annulus or volva and is generally smaller it would be difficult to confuse it with one of the white-spored Amanitas.

The *stipe* is usually 1 1/2 to 2 1/2 in. long and 3 to 5 mm thick. It is tough, white or light tan in color, and is covered with fine hairs. It lacks both volva and annulus, and the base of the stipe is not bulbous.

The distinguishing features of this species are (1) its habitat, grassy places such as lawns; however, this should never be used as the sole identifying characteristic, (2) its habit of frequently forming fairy rings, (3) the pale tan color, (4) the broadly umbonate pileus seen in some specimens, and (5) the fine appressed hairs that cover the stipe. Singer[169] reported that Canadian and English gardeners have cultivated this species for over 50 years, but Buller's[31] account of his cultivation of *M. oreades* does not indicate that it was a very high-yielding or profitable venture.

The Chantrelle (*Cantherellus cibarius*)

The Chantrelle, *Cantherellus cibarius*, is highly prized as a food item in many sections of Europe and by the knowledgeable collector in the U.S. Some canned or dried Chantrelles are imported into the U.S. from Sweden and possibly other European countries. In spite of the fact that it is white-spored, as are some of the poisonous wild mushrooms, the Chantrelle is so distinctive in form and color that it is very unlikely that another species would be mistaken for this one.

The *pileus* is yellow in color and is typically

FIGURE 24. The Fairy Ring Mushroom, *Marasmius oreades.* Note slight umbo (u) and short gills at margin (g). (Photograph by G. F. Atkinson.)

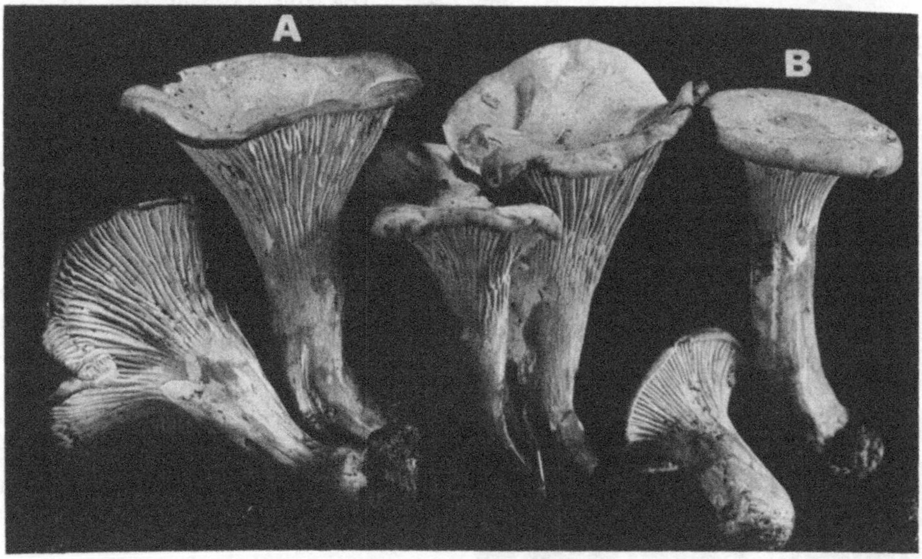

FIGURE 25. The Chantrelle, *Cantharellus cibarius.* Pileus *A* is infundibuliform; pileus *B* plane. (Photograph by G. F. Atkinson.)

infundibuliform in shape, although some specimens (see *B* in Figure 25) may have a nearly flat pileus. It varies from 1 1/4 to 4 in. in diameter. In young specimens the margin of the pileus is curved in but later it may be merely curved down or it is upraised and is frequently wavy.

The *gills* are unmistakably decurrent, narrow (2 to 4 mm) and are about 15/cm in number at the margin. They are chrome or yellow in color. They commonly anastomose or fork (see directly below *A* in Figure 25). The flesh above the gills is white and is 1 to 2 cm thick near the stipe.

The *stipe* is 1 1/4 to 2 1/2 in. long and is 6 to 15 mm thick at the top. It may be of uniform thickness or may taper downward, and the base is enlarged but not bulbous.

The Chantrelle grows on the ground in both deciduous and coniferous forests and may be found scattered or in groups. In southern Ohio I have rarely found them in groups of more than three or four. Principal distinguishing features are (1) the usually funnel-shaped pileus, (2) the yellow or chrome color, (3) the gills which are decurrent and thus extend down the stipe, and (4) the thick, frequently forked and anastomosed gills.

Christensen[37] described four other species of *Cantherellus* and listed all of them as edible, and Krieger[110] listed eight species, all but one of which were labeled as edible. It is possible that a superficial examination of the Jack-o-Lantern (*Clitocybe illudens*, sometimes called the False Chantrelle) might lead to its misidentification as a Chantrelle; however, the flesh of *C. illudens* (Figure 64) is pale yellow rather than white as is that of the Chantrelle, and the gills of the latter are blunt-edged and forked or anastomosing in contrast to the sharp-edged, unforked gills of *C. illudens*.

The Honey Mushroom (*Armillaria mellea*)

The Honey Mushroom (*Armillaria mellea*, sometimes also called the Oak Fungus or Shoestring Fungus) is an extremely common mushroom in many deciduous forests. Krieger[110] stated that it causes one of the worst tree diseases, but Christensen[37] pointed out that there is good evidence that it follows rather than causes the death of orchard and forest trees. Whether or not it is a primary pathogen, it may be collected by the bushel in wooded areas in the proper season, and may be readily dried or canned for future use. The Honey Mushroom is a favorite of Bohemians, who call it "vaclavka" and Polish-Americans, who refer to it as "putpinki" (Blei[21]).

The *pileus* is usually convex (Figure 26) but may be plane when fully expanded. Before expansion the pileus is nearly spherical and the margin is attached to the stipe by a layer of woolly mycelium which breaks as the pileus expands, leaving a veil at the margin of the pileus and an annulus on the stipe. When I have seen it, the color is usually pale wood-brown to dark reddish-brown, although it has also been described as ochre-yellow or honey colored. The surface of the pileus is commonly covered with small tufts of hairs which appear scale-like and are darker than the actual pileus surface. The surface of young specimens may be sticky or slimy, but this is not apparent in dry weather. The margin of the pileus is turned down.

The *gills* are adnate or decurrent; they are widest near the stipe (6 to 11 mm) and become gradually narrowed toward the margin. They are white or pale yellow in color and are about 18 to 20/cm in number at the margin. The flesh above the gills is firm, white in color, and may be nearly 1 in. thick near the stipe but becomes thinner as it approaches the margin.

The *spores* of the Honey Mushroom are white.

The *stipe* is 1 to 6 in. long and 1 to 2.5 cm thick. It is uniform in diameter but may taper at the base to a pointed structure; however, this pointed structure in no way resembles the long, taproot-like structure seen in *Collybia radicata*.

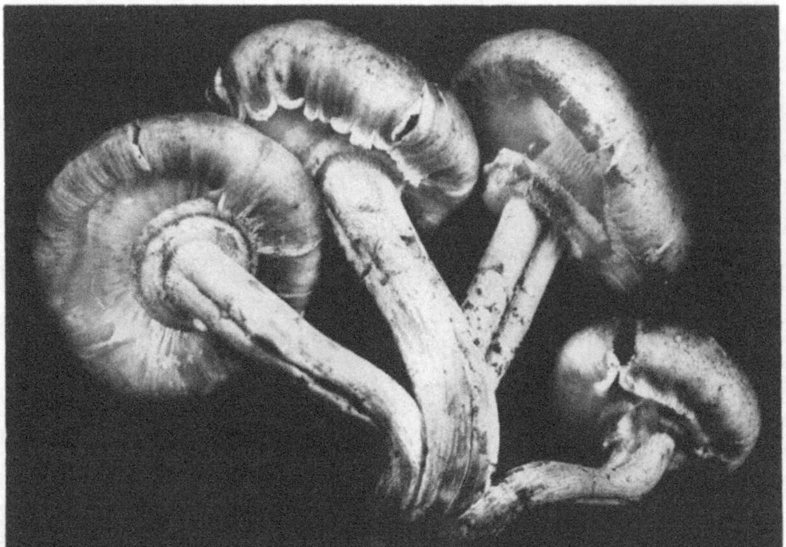

FIGURE 26. The Honey Mushroom, *Armillaria mellea.* (Photograph by G. F. Atkinson.)

The *annulus* is quite obvious when first formed but in many specimens it soon disappears and this may confuse the beginner since proper identification then becomes difficult or impossible to one who is not thoroughly familiar with the species. In view of the evanescent nature of many of the annuli, the generic name *Armillaria* seems an especially unfortunate choice since it is derived from the Latin word meaning "ring" and obviously refers to the annulus.

In areas where *A. mellea* is prevalent, evidence of its presence is often provided by the finding of extensive, dark, often flattened rhizomorphs (Figure 27), structures which probably gave rise to the common name, Shoestring Fungus. These flat, many-branched structures may be found extending a yard or more on the surface of the wood of trees or logs from which the bark has fallen away, or they may be found by pulling off pieces of the bark. The Honey Mushroom is found in late summer and autumn (August through November in northern Illinois) around dead trees or stumps. It may be solitary or in clusters, as in Figure 26. It is possible that the Jack-o-Lantern (Figure 64) might be confused with the Honey Mushroom, but the decurrent gills of *C. illudens* extend much farther down the stipe than they do in *A. mellae*; also, its color is bright orange and it has no annulus.

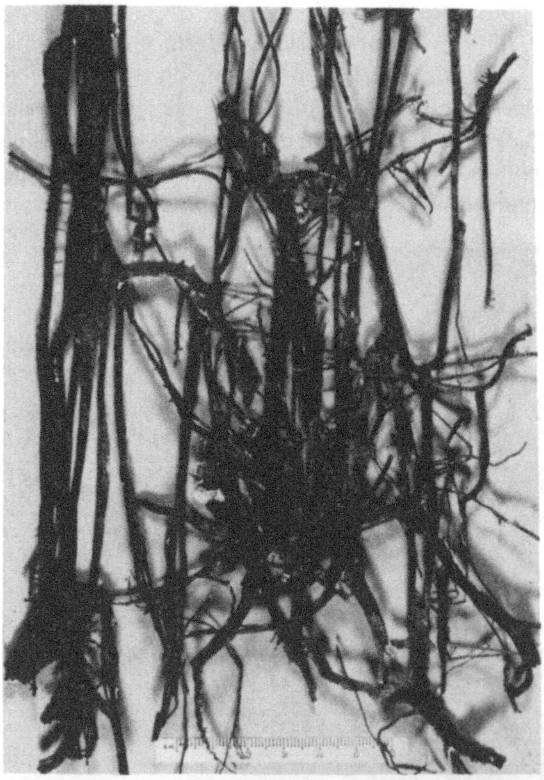

FIGURE 27. Rhizomorphs of *Armillaria mellea*. Because of its production of long, flat, black rhizomorphs, *A. mellea* is sometimes called the "shoestring fungus."

The Glistening Coprinus (*Coprinus micaceus*)

In addition to the Glistening Coprinus (*Coprinus micaceus*, see Figure 28), two other species of the genus Coprinus are recommended as edible mushrooms. There is one precautionary note: there are several reports that certain species of this genus may be poisonous if eaten with alcohol or if there is alcohol in the stomach. Therefore, until this matter is further clarified, it is recommended that if any species of *Coprinus* is eaten, alcohol should not be imbibed before, during, or immediately after the meal. *Coprinus micaceus* is the smallest of the three species described here.

The *pileus* is conic to campanulate, buff to tawny-yellow in color, and is 1 to 2 in. in diameter. The surface is often covered with small, glistening particles which present an appearance suggestive of fine mica schist, probably the origin of the common name. The pileus is striated or furrowed from the margin almost to the center.

The *gills* are narrow (about 3 to 4 mm) and number about 20/cm at the margin. They are white at first and then become tan. Then as the spores mature the gills become black and liquify, starting at the margin – a characteristic of all species of *Coprinus* (Figures 29 and 30). The flesh above the gills is quite thin as it is in all species of this genus.

The *spores* of *C. micaceus* are black.

The *stipe* of the Glistening Coprinus ranges from 1 to 4 in. in length. It is white, hollow,

FIGURE 28. The Glistening Coprinus, *Coprinus micaceus*. (Photograph by G. F. Atkinson.)

FIGURE 29. The Inky Cap, *Coprinus atramentarius*. Above, cluster of basidiocarps in their natural habitat; below, longitudinally split basidiocarps showing darkening beginning at the margin of the gills. (Photographs by G. F. Atkinson.)

brittle, and 4 to 8 mm thick. There is no trace of either annulus or volva.

This species occurs from May to November in grassy ground around the bases of trees and stumps. It occurs in large clusters several times at the same site and this may continue over a period of years, depending upon how much food is available there for the fungus. The Glistening Coprinus is a very rapidly maturing species and so should be eaten when young when the gills are white.and shortly after being collected.

Care should be taken not to confuse species of *Coprinus* with species of *Panaeolus* (Figures 69, 70, and 71) which also have black spores, since several Panaeoli are reported to be poisonous or hallucinogenic. Although the spores of *Panaeolus* are also black, the pilei and gills do not deliquesce as they do in *Coprinus*, and, as may be seen in the

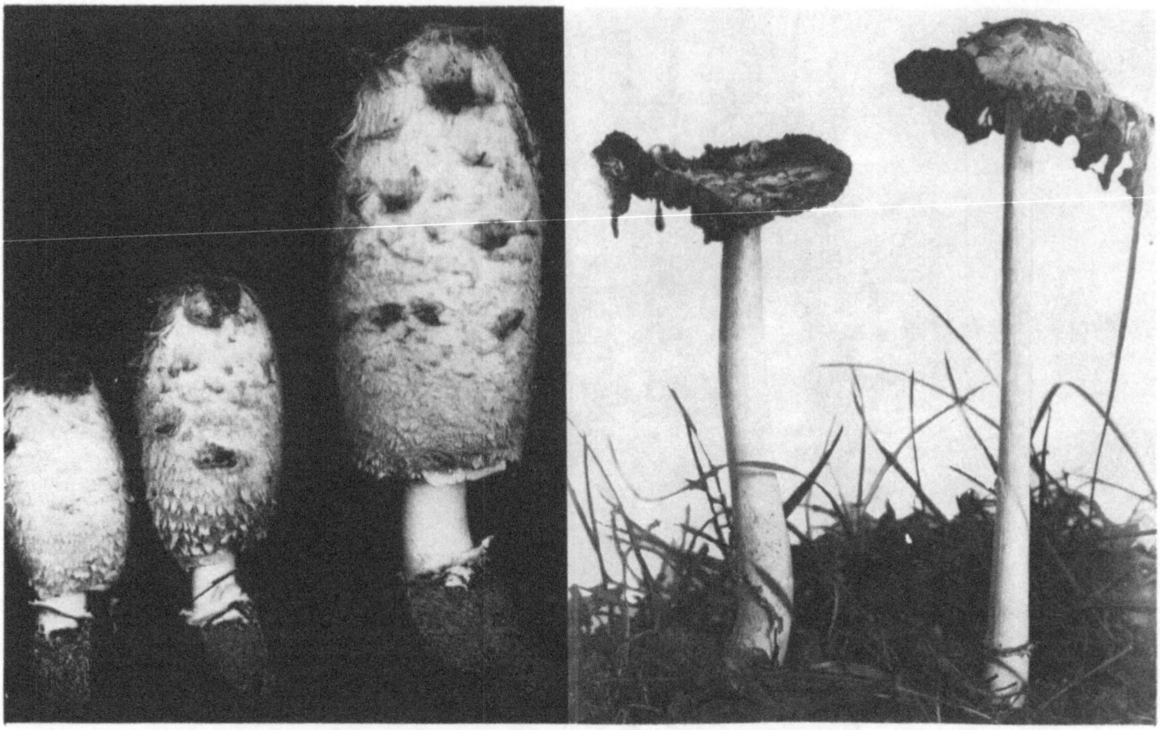

FIGURE 30. The Shaggy Mane, *Coprinus comatus.* Left, newly formed basidiocarps, about the right size for eating. Right, two basidiocarps in which the pilei and gills are nearly all liquified; note the drop of inky liquid hanging down from one pileus. (Photograph by G. F. Atkinson.)

illustration of *Panaeolus retirugis* (Figure 69), the surface of the pileus is distinctly mottled.

The Inky Cap (*Coprinus atramentarius*)

The Inky Cap (*Coprinus atramentarius*, see Figure 29) is closely related to the Glistening Coprinus but is typically a somewhat larger mushroom. The admonition regarding alcohol also applies here, since it is most frequently supposed to be involved in alcohol-mushroom poisoning.

The *pileus* is at first conic or oblong but may later become campanulate and as it expands further, splitting may occur at the margin. It is 1 to 3 in. long and has a gray surface which is darker in the central region above the stipe. The surface is striated or furrowed, is silky or mealy in appearance, and is sometimes covered with tiny scales.

The *gills* are free, rather closely spaced (20 to 30/cm at the margin) and are 8 to 15 mm wide. They are white at first but soon become black as the spores mature. Darkening begins at the margin as shown in the lower part of Figure 29. The darkening progresses toward the center, is accompanied by liquefaction, and soon the entire pileus is converted to a black, inky liquid. As in the preceding species, the flesh above the gills is very thin.

The *spores* of *C. atramentarius* are black.

The *stipe* is commonly 2 to 5 in. long but longer ones are not infrequently found; it is 8 to 15 mm thick. Christensen[37] described the stipe as being of uniform thickness but the longitudinally split specimens shown in the lower half of Figure 29 do not support this point. There is only a faint trace of an annulus appearing as a line around the stipe.

The Inky Cap is a very common species in late summer and autumn. It commonly occurs in dense clumps and may be found around hardwood stumps or in areas where there is rich soil such as open woods, lawns, or gardens.

The Shaggy Mane (*Coprinus comatus*)

The Shaggy Mane is the largest of the Coprini and, in my opinion, the most tasty. It is one of the most, perhaps the most, easily recognized of all of the gill fungi. After the collector has seen the Shaggy Mane once or twice he is never likely to confuse it with any other mushroom species, even other species of *Coprinus*. Few mushrooms should ever be first identified by use of illustrations alone, but *Coprinus comatus* (Figures 3 and 30) is one that may be safely identified in this manner. It is a fragile, quickly maturing mushroom and should be eaten shortly after being collected. Being a *Coprinus*, it would be best not to ingest it when there is alcohol in the stomach.

The *pileus* is campanulate and 1 1/4 to 3 in. long before expansion; after expansion it may attain a length of 6 in. or more. The surface of the pileus is white at first and covered with long scales which give it a shaggy appearance. Later, the color may become tan or purplish-tan but the color will be seen to be in the scales. The young, unexpanded pilei (2 to 3 in. long) are best for eating.

The *gills* are thin and very closely placed, 25 to 30/cm in number at the margin. They are 6 to 15 mm wide, at first white, then pink, and finally becoming black. Shortly after starting to become black, the gills begin to liquify at the margin (Figure 30) and finally the gills and pileus are completely reduced to a black, inky liquid. The flesh above the gills is very thin so that when the pileus is eaten the major edible portion consists of gills.

As in other Coprini, the *spores* of *C. comatus* are black.

The *stipe* is usually 3 to 6 in. long but not infrequently may be much larger, and giant specimens 12 to 14 in. tall may be seen occasionally. The stipe is hollow, 6 to 15 mm thick at the top, and gradually tapers upward. It is white and smooth or slightly scaly. An annulus is sometimes present on the stipe but it soon disappears altogether.

The Shaggy Mane is found in autumn – solitary, scattered or in clumps. It is to be sought in rich soil – lawns, fields, parks, along roadsides, etc. There are no other mushrooms with which typical specimens of this species could be confused. To prepare this mushroom for eating, the lower part of the stipe (the part extending out of the campanulate pileus) should be cut off and discarded. The pileus should then be split longitudinally. The gills may then be observed and if any blackening appears at the lower portions of the gills, these portions should be cut off and discarded. This mushroom is quite fragile and must be cooked with care. It is at its best when carefully sautéed in butter.

The Fat Pholiota (*Pholiota adiposa*)

The Fat Pholiota is an edible but not especially tasty mushroom. It is included here because it is fairly easily identified and often occurs in rather large clumps (Figure 31) – a characteristic which makes collecting for the table relatively easy.

The *pileus* is convex although in some instances (e.g., upper three basidiocarps in Figure 31) may have a small umbo. Christensen[37] listed pileus diameters as being up to nearly 5 in. but in fruitings I have seen, the size range was usually 1 to 2 1/2 in. Color ranges from pale yellow to chrome yellow and on the surface of the pileus there are concentric rings of flat, brown scales. The surface is typically slimy or sticky. Sometimes the remains of a veil may be seen on the margin.

The *gills* are adnate, yellow in color at first but later becoming dark brown as the spores mature, 7 to 12 mm wide near the stipe and 15 to 20/cm in number at the pileus margin. The flesh above the gills is thick in the region of the stipe attachment, but it thins rather abruptly toward the margin; it is pale yellow in color.

The *spores* of *Pholiota adiposa* are yellow-brown.

The *stipe* is 2 to 5 in. long and 8 to 15 mm thick at the top. It is yellow like the pileus, frequently curved, and is scaly up to the *annulus*, which quickly disappears. The stipe is hollow and seems to be especially susceptible to small insect infestation (Figure 32) so usually it should be discarded or at least split and examined.

P. adiposa occurs abundantly in autumn in clumps on logs and stumps of deciduous trees or may be found on the damaged portions (branch stubs, lightning wounds, etc.) of living trees.

The Jack-o-Lantern (*Clitocybe illudens*) might conceivably be confused with *P. adiposa*, but the gills of the former are decurrent, often extending well down the stipe, as opposed to the adnate gills of the latter. Furthermore, the surface of the pileus of *C. illudens* is not slimy or sticky, as is that of *P. adiposa*. It seems possible that *P. adiposa* also might be confused with either of two related species: *P. squarrosa* (Figure 33) and *P. squarrosoides*, the latter of which also has a viscid pileus surface but is listed as edible by Krieger.[110] Both Smith[171] and Christensen[37] listed *P. squarrosa* as edible, but Shaffer[163] reported that it is poisonous if taken with alcohol. Since the surface of the pileus of *P. squarrosa* is not viscid, and the scales are erect rather than flat, careful examination will prevent this species from being confused with *P. adiposa*.

FIGURE 31. The Fat Pholiota, *Pholiota adiposa.* (Photograph by G. F. Atkinson.)

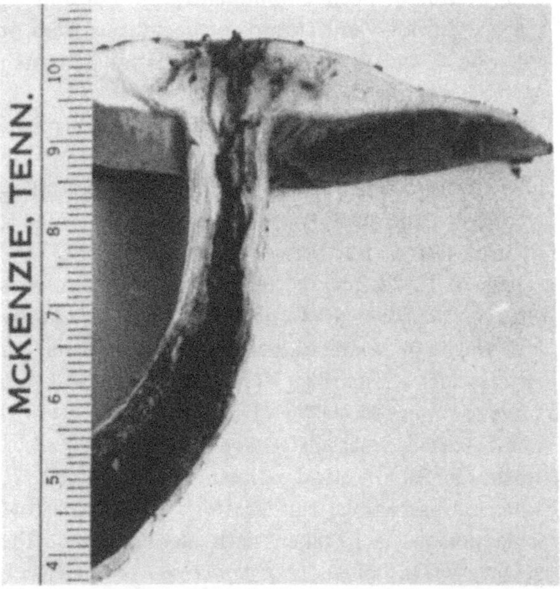

FIGURE 32. Basidiocarp of *Pholiota adiposa* split longitudinally to facilitate measuring gill width. Note extensive insect damage in center of stipe.

The Unkempt Mushroom (*Strobilomyces strobilaceus*)

This, as well as all mushrooms subsequently described in this section, is not an agaric and so does not form its spores on gills. The Unkempt Mushroom, as well as *Boletus edulis*, the species described immediately following this one, is a member of the family Boletaceae, the members of which are fleshy, and there is a layer of tubes (pores) in which the basidiospores are formed on the lower side of the pileus where the gills of an agaric would be. Ainsworth and Bisby[2] reported that there are 220 species in the family Boletaceae, of which about 200 species are in the genus *Boletus*. Many of these are difficult to identify, even when microscopic characteristics are used as an aid. For books dealing specifically with this family, the work of Coker and Beers[44] or the more recent work of Smith and Thiers[173] should be consulted.

The present species, *Strobilomyces strobilaceus*, is generally referred to as the Cone-Like Boletus

FIGURE 33. *Pholiota squarrosa.* This species might be confused with *Pholiota adiposa* by the beginner, but the upper surface of the pileus of *P. squarrosa* is not sticky. There is one report that this species is poisonous if taken with alcohol.

but here the suggestion of Krieger[110] has been adopted and it is referred to as the Unkempt Mushroom, a common name which seems especially appropriate (Figure 34). Christensen[37] stated that the genus *Strobilomyces* contains only a single species, but Krieger makes passing reference to another species, *S. camphoratus*, which he described as having a musky-camphor odor. Smith and Thiers[173] do not recognize the name *S. strobilomyces* as a valid one but place it in the synonymy of *Strobilomyces floccopus*. These latter authors also recognize another species, *S. confusus*.

There seems to be no difference of opinion regarding the edibility of this species, but there does seem to be some difference of opinion regarding its desirability as an edible species. Thus, Smith and Thiers[173] stated that it has little to recommend it for the table, Christensen[37] merely listed it as being edible, but Krieger[110] stated that it is highly prized among Bohemian people. Obviously, tastes among people do vary! It is included here because it can be easily recognized and so can be used by those individuals who may find its taste acceptable or appealing.

The *pileus* is hemispherical or convex when expanded and ranges in diameter from 2 to 4 in. The surface is covered with dark brown or almost black tufts of hyphae which often appear pyramidal in shape. The surface under the scaly tufts is pale brown in color.

The *pores* are 1.5 to 2.5 cm long and are soft and dark brown. The flesh is 1 to 2 cm thick near the stipe but disappears near the margin.

The *spores* are dark brown in color.

The *stipe* is about 2 to 5 in. long and 1 to 2 cm thick; it may be of uniform thickness or may taper upward. It is easily detached from the pileus and may be covered with scales like those on the pileus.

S. strobilaceus grows scattered in deciduous woods or at the edges of such woods from July through September. The principal distinguishing features are (1) the shaggy, unkempt appearance which is due to the presence of large, prominent tufts on the surface of the pileus (other members of the Boletaceae have pilei which are smooth, scaly, or mucilaginous, but not covered with large tufts), and (2) the presence of a layer of pores rather than gills.

The Select Boletus (*Boletus edulis*)

As noted in the discussion of the Unkempt Mushroom, the genus *Boletus* is a very large one. Unfortunately, many of the species can be identified only with the aid of a microscope, and Christensen[37] stated that many of them are not readily identifiable. The Select Boletus, *Boletus edulis* (Figure 35), is included here because it is considered by many people to be one of the most delicious of mushrooms. Thus, Smith and Thiers[173] stated that it is one of the best edible

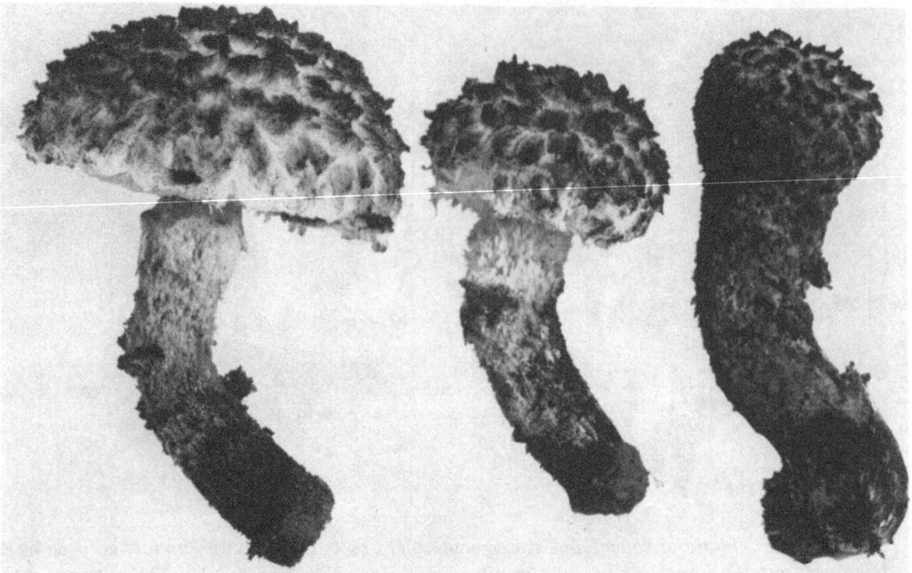

FIGURE 34. The Unkempt Mushroom, *Strobilomyces strobilaceus.* (Photograph by G. F. Atkinson.)

FIGURE 35. The Select Boletus, *Boletus edulis.* (Photograph by G. F. Atkinson.)

fungi in Michigan. These authors consider it a "collective" species comprised of four varieties. This is a much prized mushroom in Europe, is widely collected and marketed there both dried and canned, and appears in various countries there under a variety of names:

cèpe	France
steinpilz	Germany and Switzerland
herrenpilz	Austria
porcino	Italy
baroviki	Poland
balya-grib	Russia
stensopp	Denmark

In the Introduction, it was noted that there may be some difference of opinion as regards what should be considered a mushroom. Oddly enough, this difference of opinion ended up in a magistrate's court in Chichester, England, where it was questioned if flavoring made from *B. edulis* could properly be called mushroom flavoring. There it was determined (Anonymous[5,6]) that *B. edulis* is a mushroom even though it is not a gill fungus; hence, in England at least, a fleshy fungus legally may be a mushroom even if it is not an agaric!

The *pileus* of the Select Boletus is convex (or sometimes almost plane) and usually ranges in size from 3 to 6 in. The surface is smooth, moist (sometimes slightly sticky), and in color may be grayish-red, brownish-red, or tawny-brown, and is usually paler at the margin.

The *tubes* are white or pallid and become greenish-yellow with age. The tube layer is 1.5 to 2 cm thick and can be easily peeled from the pileus. The tube layer does not have a plane surface but curves up near the stipe. The flesh above the pore layer is 1 to 2 cm thick and is white (Christensen[37]) or white or yellowish (Krieger[110]). The tube layer is removed before cooking.

The *spores* are variously described (depending upon variety) as olive-brown, olive-ochraceus, olivaceus, or olive-black.

The *stipe* is 2 to 6 in. long, stout (1 to 2 cm in diameter), and is nearly white or is colored like the pileus. It usually occurs in woods in July and August.

FIGURE 36. Fleshy fungi may sometimes be parasitized by other fleshy fungi. Here *Boletus parasiticus* is parasitizing the Black-Centered Puffball, *Scleroderma vulgare.* Smith and Thiers[173] do not state if *B. parasiticus* is edible or poisonous, but because of its close association with *S. vulgare* which Zeitlmayr[204] states is poisonous, it would be best not to attempt to eat this species until more information is at hand.

Although most species of *Boletus* are identified with difficulty, there is one unmistakable species, *B. parasiticus*. Identification of this species is made easy because it parasitizes *Scleroderma vulgare* (Figure 36). This black-centered puffball is unmistakable and if a bolete is found attached to it, one may be sure that the bolete is *B. parasiticus*. McIlvaine[123] stated that this parasitic bolete is edible, but since it is intimately associated with *S. vulgare*, which Zeitlmayr[204] stated is poisonous, it might be best to leave *Boletus parasiticus* alone.

A smaller, related genus is *Boletinus*, which is easily distinguished from *Boletus*. The pores of *Boletinus* are so shallow that they can scarcely be called tubes, and this "pore" layer is not easily peeled from the pileus, as is the tube layer of *Boletus*. The pore walls of *Boletinus* are made up of thick, irregular veins. This characteristic and the shallow pores are shown in Figure 37, which illustrates *B. porosus*, an edible (but not especially tasty) species which has a lateral or eccentric stipe.

The Sulphur Mushroom [*Polyporus (Laetiporus) sulphureus*]

Besides the Boletaceae which includes species that form pores, there is another large family of fungi whose members also produce their basidiospores in pores. This second family is the Polyporaceae, which includes those pore fungi in which the reproductive structures are, for the most part, tough, leathery, or woody. Some of these structures are annual while others are perennial and hence get larger from year to year; only a few of the annual species are suitable for eating. Unlike

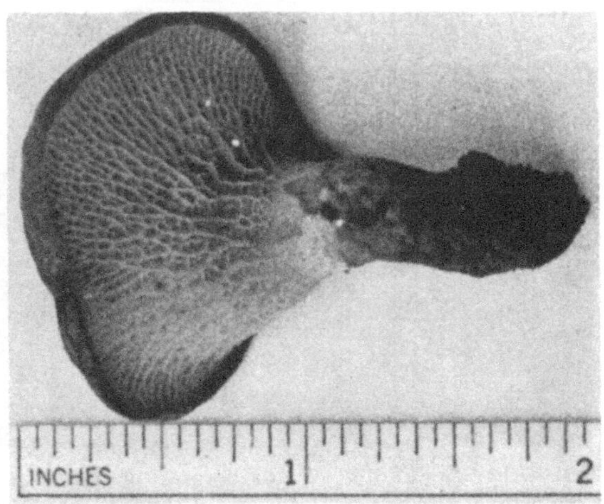

FIGURE 37. Lower surface of *Boletinus* sp. Note shallow pores instead of deep tubes as in *Boletus* or gills as in the agarics.

the boletes, the tube layers of the polypores cannot be easily separated from the pileus. For those who may wish to learn more about this family, the work of Overholts[139] or Lowe[119] is recommended. The former describes 235 species, the latter, 146 species.

Very few species of Polyporaceae are fleshy enough to be considered suitable for eating. The Sulphur Mushroom, *Polyporus sulphureus* (more recently termed *Laetiporus sulphureus*), is one such exception.

The *pileus* of the Sulphur Mushroom grows shelf-like and with no stipe from woody substrates in large groups (Figure 38). Pilei are usually 2 to 6 in. in diameter, but Christensen[37] reported that they may be as large as 30 cm (12 in.) in diameter. When young, they may be bright orange, reddish-orange, or sulphur-yellow in color but they fade to a pale tan with age. Large pilei of this species are usually somewhat tough and so with larger specimens it is best to use only the more tender margins of the pilei.

The *pores* on the lower surfaces of the pilei are small, shallow, and often difficult to see without a lens. The pore layer is yellow when fresh. The flesh above the pore layer is commonly no more than 1/2 in. thick and it is white or pale yellow in color. When fresh the pileus is firm but breaks easily.

This is a common species from July to September on stumps, logs, or dead areas on the trunks of living deciduous trees. Apparently it occurs only rarely on coniferous trees. The principal distinguishing features are (1) the unique sulphur-yellow or orange color, (2) the minute, often difficult-to-see pores, and (3) the non-stipitate pilei growing shelf-like in large groups on woody substrates.

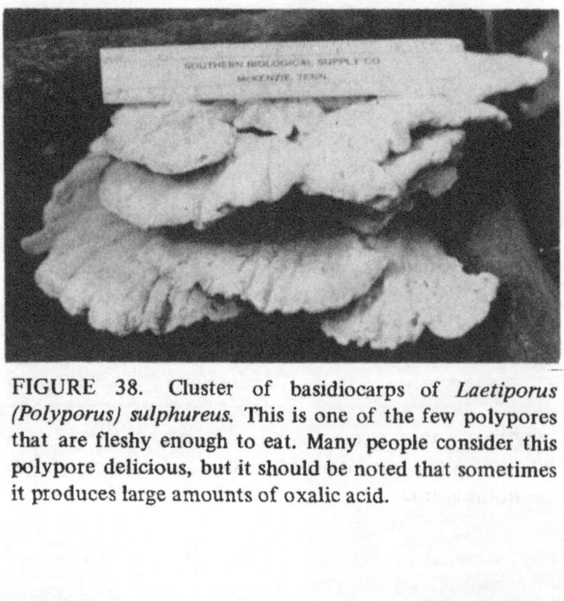

FIGURE 38. Cluster of basidiocarps of *Laetiporus (Polyporus) sulphureus*. This is one of the few polypores that are fleshy enough to eat. Many people consider this polypore delicious, but it should be noted that sometimes it produces large amounts of oxalic acid.

Krieger[110] merely described the Sulphur Mushroom as edible, while Christensen[37] stated that it is often delicious when cut in thin slices and fried. I know many individuals who consider this species a very desirable one. However, because of its production of relatively large amounts of oxalic acid, I prefer not to eat it.

Chicken of the Woods (*Polyporus frondosus*)

Chicken of the Woods (or Hen of the Woods) is another species of Polyporaceae which is not too woody or too leathery to eat. Christensen[37] said nothing about its edibility and Krieger[110] merely stated that it is edible, but in certain areas of Illinois there are individuals who consider this species a delicacy. The fruiting body of *Polyporus frondosus* is a compound one – that is, it is made up of many basidiocarps (20 or more) growing very close together (Figure 39) from a common point and hence forming a large mass. Such masses which I have collected usually have been about 8 to 10 in. in diameter, but Christensen noted that they may be as large as 20 in. Presumably, the common name derives from the rather vague resemblance of such a mass to a chicken or a hen with her chicks clustered around her.

The *pileus* of a single basidiocarp of a cluster is 3 to 10 cm wide. The color may be smoky-gray or tan to brown. The surface of the pileus appears hairy. The margin may be straight or wavy, and the pileus may be thick or thin.

The *pores* are 1 to 3 mm long, and the pore surface is white.

The *stipe*, attached laterally, is 1 to 3 cm thick, and 1 to 4 cm long. The stipes are intergrown and usually the cluster appears to have one very large "stem" from which arise many short, branched stipes.

This species is found in autumn and, while it does not occur frequently, can scarcely be classed as rare. Krieger[110] stated that it grows from the bases of living oaks, but I have never seen it except on the ground in deciduous woods, often near stumps. In northern Illinois, *P. frondosus* is sought by Bohemians who refer to it as "kotrc" (Blei[21]).

Chicken of the Woods is sometimes considered to be too leathery to be a very desirable edible species, but because of its firm consistency it is a good species for pickling. The principal distinguishing features are (1) the smoky-gray to

FIGURE 39. Chicken of the Woods, *Polyporous frondosus.* (Photograph by G. F. Atkinson.)

smoky-tan or brown pileus with white under surface, (2) the short, laterally placed stipes, and (3) the habit of growth which leads to a compound structure consisting of many separate basidiocarps.

The Coral Fungi (*Clavaria* spp.)

Like the Agaricaceae, Boletaceae, and Polyporaceae, the Clavariaceae (Coral Fungi) are Basidiomycetes, but their spores are not formed on gills or in pores. On the contrary, the basidiospores of these fungi are formed on the surfaces of many branched structures which resemble coral. There are about 200 species in the family and many of them can be identified only by resorting to the use of a microscope. For the person who may be specifically interested in the Coral Fungi, the work of Coker[43] is recommended.

Christensen[37] described six species of *Clavaria* in his manual and keys them out using only macroscopic characters. Of these he listed three species as edible ones. Krieger[110] also described six species and listed five of them as edible. One of the edible species, *C. pyxidata*, is shown in Figure 40. Korf and Burdsall[108] stated that there are some poisonous species in the genus that can cause violent illness. For this reason it is recommended that the amateur not attempt to eat any Coral Fungi unless he becomes well-acquainted with the group.

The Skull-Shaped Puffball (*Calvatia craniiformis*)

All species described earlier in this section belong to families of Basidiomycetes in which the basidiospores are produced in such a manner that they are exposed on the surfaces of gills, lining tubes which open to the outside, or on the surfaces of many branched coral-like structures as in the Clavariaceae. On the contrary, the Puffballs, while they are also Basidiomycetes, produce their spores inside fruiting bodies which remain closed, or open only after the spores are mature. Such fungi are classed as Gastromycetes. There are several families of Gastromycetes but the only one with which we will be concerned here as a source of edible fungi is the Lycoperdaceae, the family of true puffballs. The two genera of this family in which we are interested are *Calvatia*, in which the wall (peridium) of the upper part of the fruiting body breaks up and disappears at maturity, and *Lycoperdon*, in which a pore forms at the top at maturity. For more information regarding the Gastromycetes, the reader is referred to the works of Coker and Couch[45] and Kambly and Lee.[105]

The Skull-Shaped Puffball, *Calvatia craniiformis*, is a commonly encountered species and is of a very convenient size to prepare for eating (Figure 41). The more typically shaped fruiting bodies are shaped roughly like a skull – a characteristic which gave rise to the common name. Thus it has a rounded or spherical upper portion which merges into a narrower base. It is

FIGURE 40. A Coral Fungus, *Clavaria pyxidata*. This is an edible species, but since some Coral Fungi have been reported to be poisonous, they should not be eaten unless the collector is sure as to species identification. (Photograph by G. F. Atkinson.)

FIGURE 41. The Skull-Shaped Puffball, *Calvatia craniiformis*, which appears in late summer and autumn. (From Gray.[79] By permission of Holt, Rinehart and Winston.)

inside of the upper part that spores will be formed, the lower stalk-like part consisting of a mass of loosely packed, spongy-appearing sterile tissue. The diameter of the upper, expanded portion may be as great as 6 in., but it is more common to find specimens that are 2 1/2 to 3 1/2 in. in their greatest diameter. The color is light tan or grayish and the surface is smooth at first, but it may later become minutely scaly and mottled in appearance. The wall, which consists of two layers (peridia), can be peeled off rather readily, exposing the snow-white inner tissue in specimens that are fresh. As the spores in this inner tissue begin to mature the tissue darkens, becomes drier, and at maturity becomes a dusty, greenish-yellow.

C. craniiformis, like other true puffballs, should be eaten only when the tissue of the entire upper portion is white in color. In *A* and *B* of Figure 42 are shown two specimens which are quite similar in external appearance except for minor differences in shape. However, when these are split longitudinally (see *a* and *b* in Figure 42), the upper

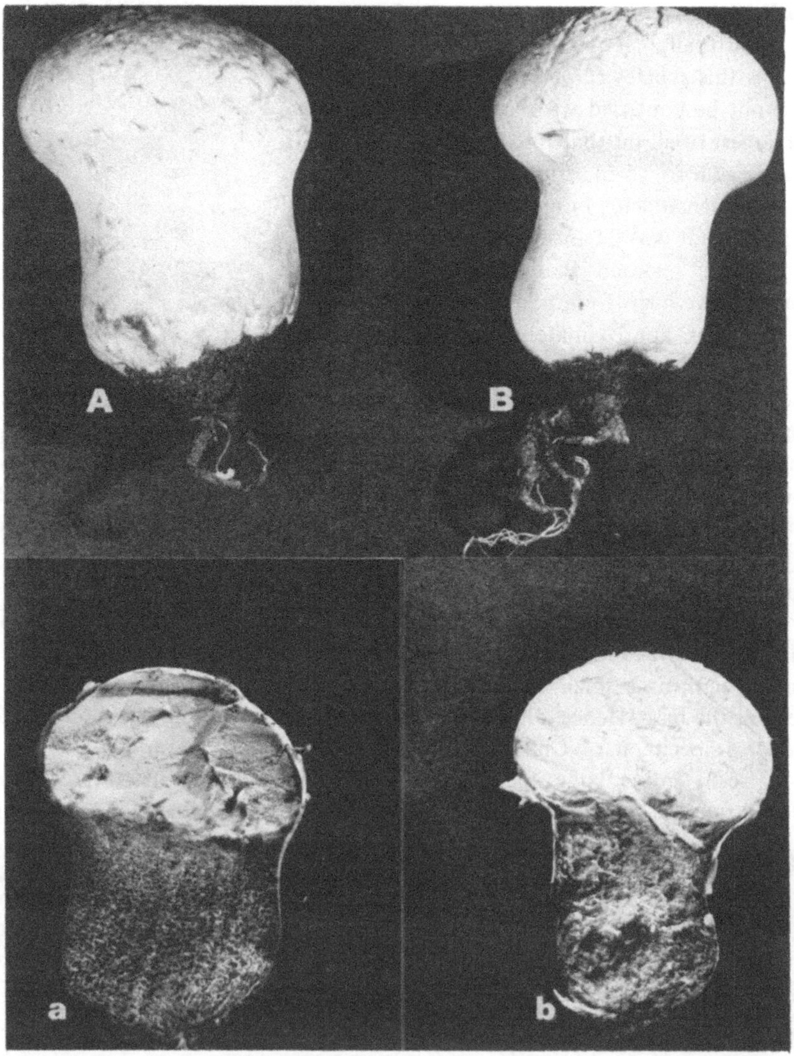

FIGURE 42. *A* and *B* – basidiocarps of the Skull-Shaped Puffball, *Calvatia craniiformis; a*, longitudinal section through *A; b*, longitudinal section through *B*. The upper spore-bearing portion of *A* has started to turn yellow and should not be eaten. The upper part of *B* is white and may be eaten. In neither case is the spongy, lower, stalk-like portion eaten.

part of *A* is seen to be darker than that of *B*, which is still white. Thus, the upper portion of *B* may be eaten, while that of *A* should not. The lower portions of both *A* and *B* have darkened; however, these portions because of their loose, spongy texture should be discarded even if they have not changed color. When peeled, sliced, and cut into smaller segments, the upper portion of *B* appears as in Figure 43.

The True Puffballs typically appear in late summer and autumn and *C. craniiformis* is no exception. It is to be sought in grassy places and open woods.

FIGURE 43. Slices of peeled *Calvatia craniiformis* ready for cooking. If dipped in egg batter, rolled in cracker crumbs, salted, and fried, these make tasty hors d'oeuvres.

The Giant Puffball (*Calvatia gigantea*)

The Giant Puffball, *Calvatia gigantea* (Christensen[37] calls this species *C. maxima*), is so unique that it cannot be confused with any other fungus. It is the largest of all puffballs. Krieger[110] listed the size range as 8 to 15 in., in which range the majority of specimens occur, but in Figure 44 is shown a specimen that was 21 in. in its largest dimension. Christensen[37] stated that specimens weighing 45 pounds have been found and that he had seen one that was approximately 2 ft in diameter.

C. gigantea does not have a stalk-like portion such as that seen in *C. craniiformis*. It is often roughly spherical although it may vary considerably from the spherical shape. The term Gastromycete means "stomach fungus" and it was probably from just such a specimen as that shown in Figure 44 that this name for the puffballs was originally derived. Having no stalk-like portion, there is usually no sterile tissue (Figure 45), although in some specimens a small amount may sometimes be seen at the base. Hence, almost all of the inner tissue of a specimen of Giant Puffball may be eaten if fresh, and a large specimen is sufficient for several families.

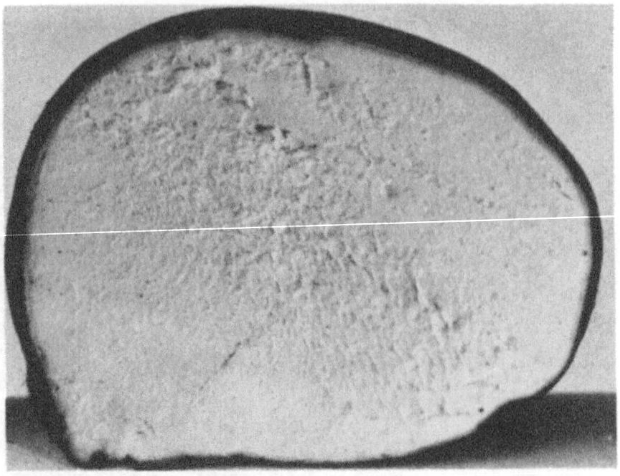

FIGURE 45. Section through small basidiocarp of the Giant Puffball showing white, homogeneous flesh inside the peridium. (From Groves.[83] By permission of the Canada Department of Agriculture.)

The Giant Puffball is found in late summer and early autumn in grassy places.

It has been noted earlier that only white-fleshed puffballs should be eaten, and then only when fresh. In Figure 46 is shown a common black-centered puffball (*Scleroderma vulgare*) of the type which should not be eaten.

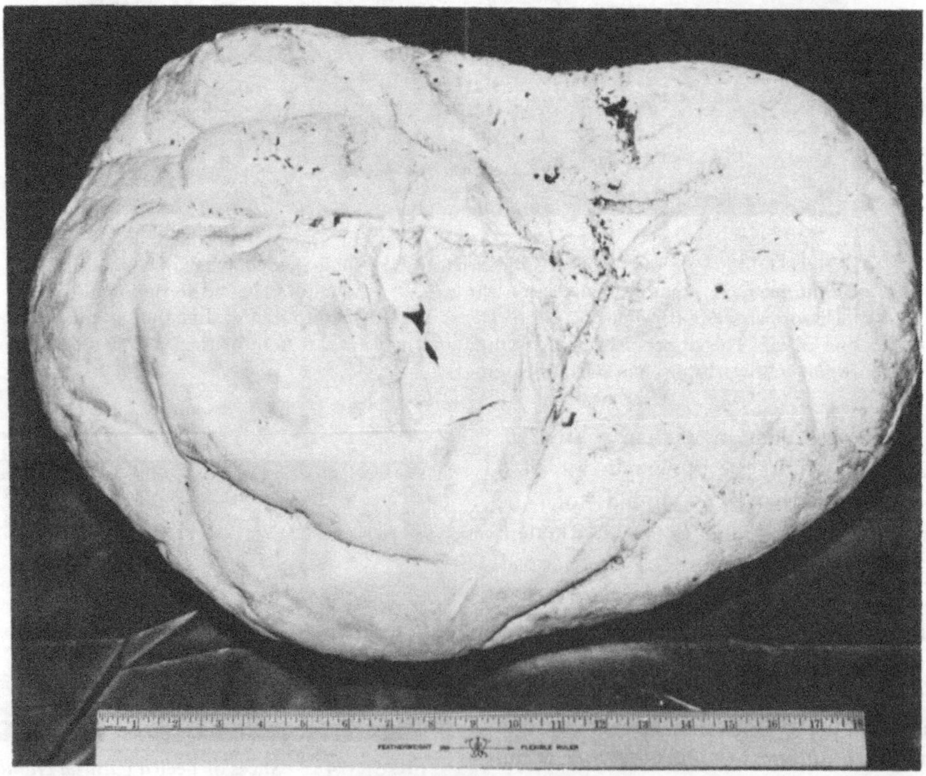

FIGURE 44. The Giant Puffball, *Calvatia gigantea*. Length, 21 in., width, 14 in., height, 14 in., weight, 16 lb., 7 oz. (Courtesy of the Union Fork and Hoe Company.)

FIGURE 46. Basidiocarps of the common Black-Centered Puffball, *Scleroderma vulgare*, about one half natural size. The basidiocarp at lower right has been broken to reveal the black center.

The Smaller Puffballs

There are two other rather common edible puffballs which should be briefly discussed here. Both are much smaller than *Calvatia craniiformis*, which in turn is smaller than *C. gigantea*. These are both members of the genus *Lycoperdon*, in which the peridium of the upper part does not break irregularly but opens by means of a pore at maturity (see *p* in Figure 47). These are the Gem-Studded Puffball, *Lycoperdon gemmatum* (Figures 47 and 48) and the Pear-Shaped Puffball, *Lycoperdon pyriforme* (Figure 48). Their typical habitat is old, very rotten wood, but both may occasionally be found on the ground. Both are roughly pear-shaped, somewhat elongated, and like the larger *Calvatia craniiformis*, both have a narrower stalk-like basal portion. Both are about 1 to 2 in. tall with a lesser diameter, and both are often found in dense clumps numbering 50 to 100 or more. Both species are typically white in color at first and on both there are short spines and warts on top (usually longer in *L. gemmatum*). The spines fall off as they mature and the surfaces become some shade of tan or brown. The peridium of the Pear-Shaped Puffball is rather tough and by the time it is removed, there is not much edible material left. However, the peridia of young, fresh Gem-Studded Puffballs are somewhat more tender, so this species does not require peeling. Therefore, if a large collection of this latter species is made, the points of attachment to the substrate may be trimmed off and the spines at the top easily washed off in running water. They may then be pickled whole (see section on The Preparation of Mushrooms) and make a unique appearing and rather tasty mushroom pickle.

The Morels (*Morchella* spp.)

The Morels (*Morchella* spp.) are representative of an entirely different class of fungi than the other edible fungi which have just been discussed. These mushrooms belong to the class Ascomycetes in which the spores (*ascospores*) are produced (usually 8 in number) in small sacs (*asci*). In the Morels the asci are arranged side by side in palisade fashion all over the surface of the upper, fertile portion of the fruiting body (*ascocarp*). In my opinion these are the most delicious mushrooms to be found growing in the U.S. Unfortunately, no one as yet has been able to bring them into

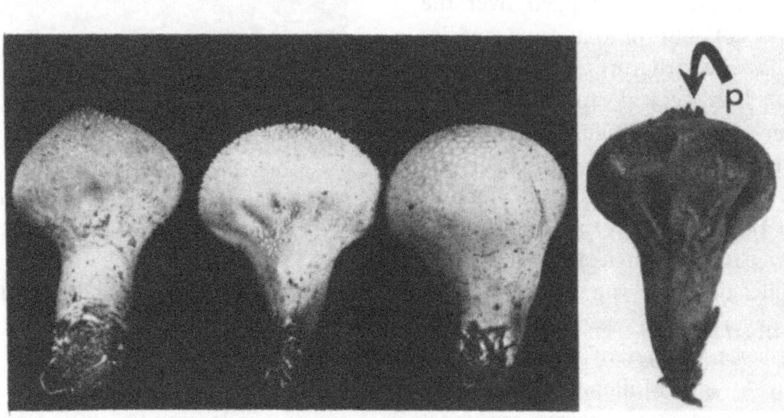

FIGURE 47. The Gem-Studded Puffball, *Lycoperdon gemmatum*. Puffball at extreme right is mature and a pore (*p*) has formed at the top, a characteristic of the genus *Lycoperdon*. (Photograph by G. F. Atkinson.)

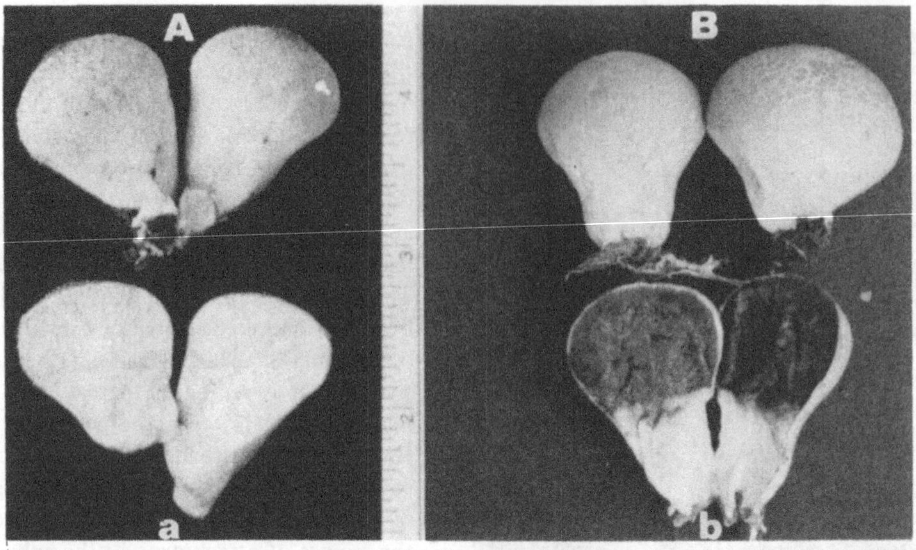

FIGURE 48. Small species of edible puffballs: *A*, basidiocarps of the Gem-Studded Puffball, *Lycoperdon gemmatum; a*, longitudinally split basidiocarp showing white interior; *B*, basidiocarps of the Pear-Shaped Puffball, *Lycoperdon pyriforme; b*, longitudinally split basidiocarp showing dark interior of the upper spore-bearing region.

cultivation and they appear only during a rather limited period in early spring, although exceptional appearances may occur, as is illustrated in Figure 49. Some canned Morels are imported into the U.S. from France. They are rather high in price (ca. $26.00 a pound) although markedly cheaper than canned truffles (Figure 50). At one time in this country they were collected in great abundance in season in Michigan, frozen, and marketed. However, the taste and consistency of Morels which have been frozen and thawed do not compare at all favorably with the fresh mushrooms or even those that have been dried and then reconstituted for eating.

A Morel ascocarp has an easily recognizable stipe and pileus (the asci are arranged over the surface of the pileus), but in only one species *(Morchella hybrida)* are structures formed which at all resemble the typical umbrella-shaped basidiocarps of an agaric. The pilei of all Morels are pitted and ridged and differ in color from the white or nearly white stipe. A genus which is closely related to *Morchella* is *Verpa*, but the two genera are usually easily distinguished from each other since the pilei of *Verpa* spp. are smooth or merely ridged, never pitted as in *Morchella*. Krieger[110] listed seven species of *Morchella*, and both Christensen[37] and Mullendore[131] listed four. In my experience four species are most commonly encountered in quantity; however, the following short key, which is based upon an unpublished key prepared by myself and the late W. G. Stover, includes six species:

FIGURE 49. Morels are rarely found except in rich soil (mostly under trees) in the early spring. In this instance Morels developed on a broken basement floor on July 12, 1964. (Photograph by Daniel C. Brautigan.)

1. Lower half of the pileus free from the stipe; more longitudinally furrowed or ridged than pitted . . . . *M. hybrida*
1. Whole pileus attached to the stipe leaving no

FIGURE 50. Canned truffles from France. This small can, which contains only 25 grams, cost $2.75 in Minneapolis, a rate of $49.77 per pound.

free margin; pileus with irregularly rounded or elongated pits 2
2. Pileus small (2 to 3 cm), gray; pits narrow, elongated . . . . *M. deliciosa*
2. Pileus usually longer than 4 cm; pits broad. 3
3. Edges of pileus ridges thick, obtuse; pits small but deep; stipe not conspicuously enlarged at the base 4
3. Edges of pileus ridges thin, almost knife-blade-like 5
4. Pileus mostly globose to oval, rounded at top . . . . *M. esculenta*
4. Pileus mostly conical in shape, pointed at the top . . . . *M. conica*
5. Ridges becoming smoky-brown, almost black at the edge, much darker than the pits; pileus mostly conical, pointed . . . . *M. angusticeps*
5. Stipe often considerably enlarged at the base, edges of ridges thin; ridges not darker in color than the interior of the pits; pits large (to 1 cm) and shallow . *M. crassipes*

Of the six species listed in the above key, there is some question as to the validity of one species. The so-called Conical Morel, *Morchella conica* (Figure 51), is, in my opinion, merely a more or less pointed form of *M. esculenta,* since I have found the typical rounded-top *M. esculenta* and

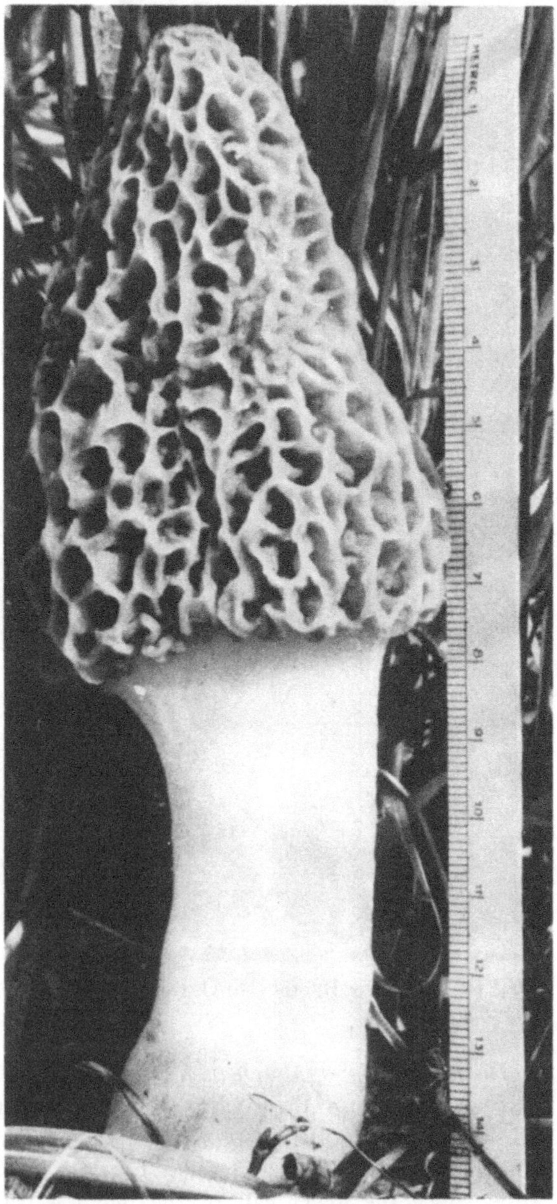

FIGURE 51. Large ascocarp of *Morchella conica.* This species is very similar to *M. esculenta* except for the cone-shaped pileus and may simply be a variant of the latter species.

the pointed-top *M. conica* growing in close proximity with various intermediate forms closely associated with them. The smallest of the six species listed, *M. deliciosa,* is easily distinguished because of its small size and its pastel-gray-colored pileus, but in my experience only a few ascocarps of this species are usually found. For these reasons only the remaining four species of the key will be illustrated and briefly described.

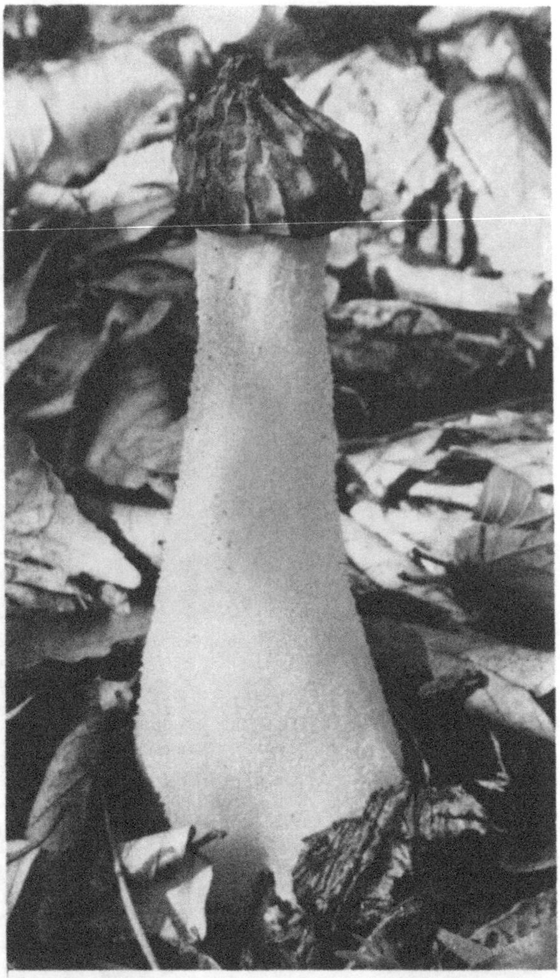

FIGURE 52. The Hybrid Morel (Spring Mushroom), *Morchella hybrida.*

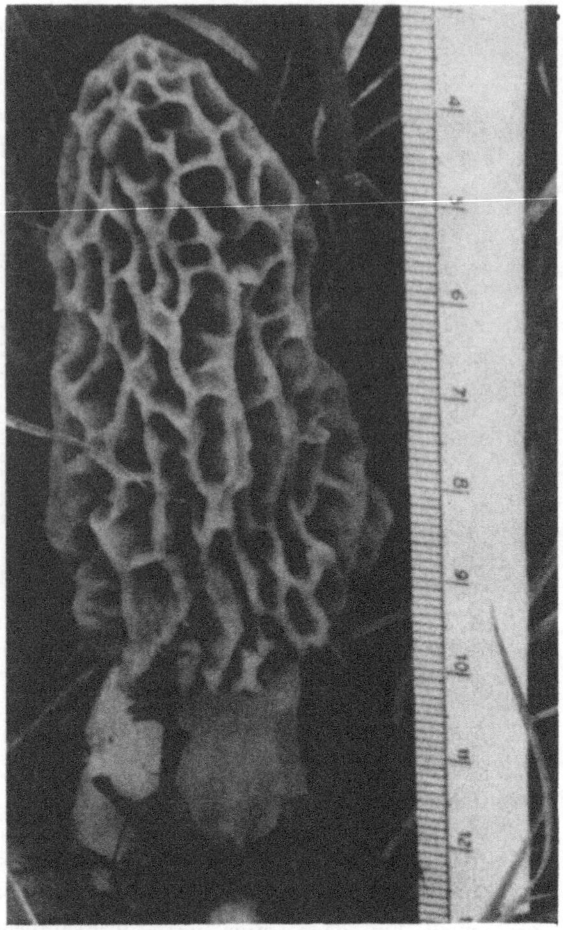

FIGURE 53. The Sponge Mushroom (White Morel), *Morchella esculenta.*

1. The Hybrid Morel (Morchella hybrida)

The ascocarp of the Hybrid Morel or Spring Mushroom (Figure 52) is typically 3 to 4 in. in total height, with the pileus small in relation to the stipe. Since the lower half of the pileus is free from the stipe and the pileus is often broadly cone-shaped, ascocarps of this species may superficially resemble an agaric with a small conic pileus and a large bulky stipe. However, the absence of gills on the bottom of the pileus and the longitudinal furrows or ridges on the upper surface of the pileus should be sufficient to avoid an error of this type. The stipe is usually snow-white in color, hollow, brittle, and often enlarged at the base. It is possible that this species might be confused with *Verpa bohemica.* However, in *V. bohemica* more of the pileus is free and in my experience the stipes are not enlarged at the base. Nonetheless, since *V. bohemica* has been reported to cause symptoms resembling intoxication, the beginner is advised not to attempt to eat *M. hybrida* until he is quite sure that he can distinguish it from *V. bohemica.*

2. The Sponge Mushroom (Morchella esculenta)

The Sponge Mushroom (Figure 53) is also called the "White Morel," although the pileus color varies from light cream to yellowish-brown. The total height of the ascocarp is commonly 1 and 3/4 to 3 in., with the pileus usually being slightly longer than the stipe. The ridges are rounded at their edges and the pits are rounded or irregular, 5 to 10 mm wide and 5 to 10 mm deep. The stipe is white in color and is hollow and fragile. Ascocarps of this species may be readily dried for future use and reconstituted immediately before cooking by soaking in water.

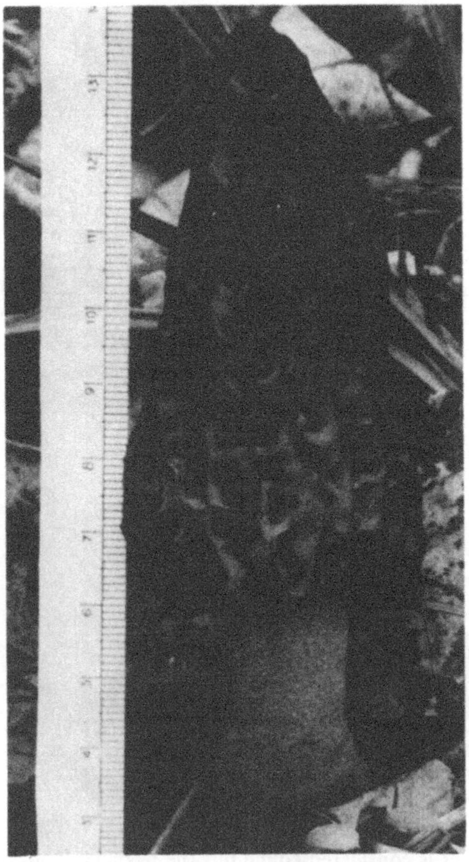

FIGURE 54. The Black Morel, *Morchella angusticeps.* Note the dark, almost black color of the ridges which separate the pits.

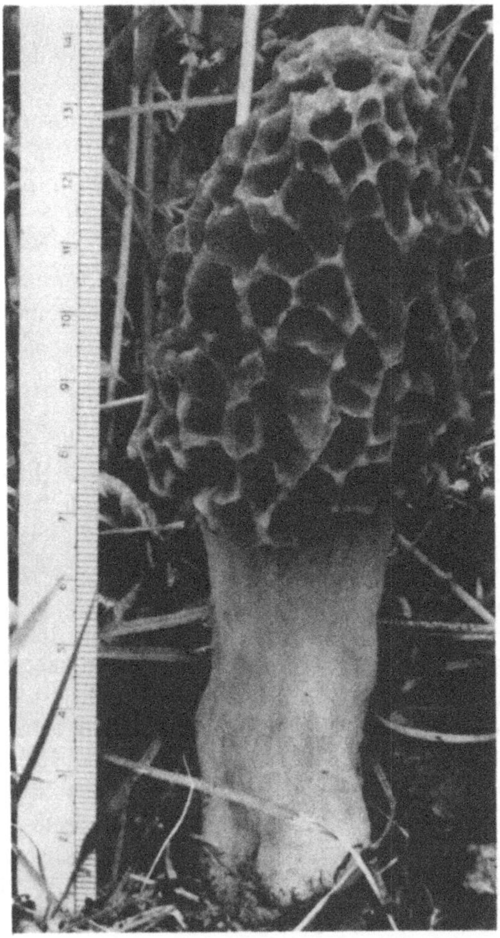

FIGURE 55. The Thick-Footed Morel, *Morchella crassipes.*

3. The Black Morel (Morchella angusticeps)

Mullendore[131] stated that the Black Morel (Figure 54) is small but edible; however, in fruitings seen by me, ascocarps usually range in total height from 3 to 5 in. and specimens up to 14 in. tall have often been found. Mullendore's colored drawing of this species approximates those ascocarps usually seen by me, but Christensen's[37] illustration labeled *Morchella conica* (Christensen, page 219) appears to be a rather typical pair of ascocarps of *Morchella angusticeps.* The stipe of *M. angusticeps* is white and hollow, but it is the pileus of this species which makes it easy to distinguish. The pileus is commonly pointed at the top, and the ridges are quite thin at their edges as compared with the rounded edges of *M. esculenta.* Much more unique, however, is the coloring of the ridge edges; they are smoky-brown, almost black, and contrast strongly with the lighter color of the pits.

4. The Thick-Footed Morel (Morchella crassipes)

Although Christensen[37] listed the height of the Thick-Footed Morel (Figure 55) as 5 to 8 cm, the present writer has found it typically to be somewhat larger, and Mullendore[131] referred to it as the "giant" among Morels. The stipe is hollow, white to flesh-colored or pale tan in color, sometimes grooved and often enlarged at the top and base. The pileus may be narrowly conical but more commonly is irregularly cylindrical and is tan or pale brown in color. The pits are large and shallow. As in *M. angusticeps* the ridges of *M. crassipes* are knifeblade-like at their edges but unlike *M. angusticeps* the ridge edges are not markedly darker in color than the pits.

Although the False Morels will be illustrated and discussed in greater detail in the section on poisonous mushrooms, a note of caution should be inserted here since at least one of the False Morels

(Gyromitra esculenta) is known to be poisonous and has resulted in a considerable number of fatalities. The poisonous *G. esculenta* is quite commonly much larger and bulkier than the true Morels (Figure 67). Furthermore, the stipes are thicker and rugose, and the pilei are typically reddish-brown in color. A close relative of *G. esculenta, Gyromitra brunnea* (Figure 56), is considered to be edible, but it is not recommended here as an edible species. Although some individuals seem to be able to eat *G. esculenta* with impunity, the beginning mycophagist should avoid it as well as its relatives. It is suggested that before any Morel is eaten by the beginner, he should familiarize himself with the False Morels (Figures 56 and 67) since they occur in the same habitats and during the same season as the true Morels.

Summation

In the foregoing discussion more than 25 different species of edible fleshy fungi have been described and illustrated. If the mycophagist learns to recognize most of these species he can expect to have mushrooms to eat the year around, although during the winter months he will have to depend upon the Velvet-Stemmed Mushroom and occasional collections of the Oyster Mushroom if there are several days of mild weather. This discussion is designed only to supplement, not to replace a good mushroom manual, and until the collector is thoroughly familiar with a species he should never consider it to be identified on the basis of illustrations alone. The more he collects and the more he uses his manual, the wider his acquaintance with mushrooms will become and he may well find that there are many additional species which are equal in taste to the small number of species discussed above. As he gains in confidence he should keep the following thought foremost in his mind: there are many old mycophagists and there are many bold mycophagists but there are not many old, bold mycophagists.

FUNGI AS ANIMAL FOOD

Most mushroom collectors encounter evidence, in the form of mushrooms that have been partially eaten, that animals of one type of another not infrequently use fleshy fungi as food. This is often of some annoyance to the mycophagist, but it also shows that organisms other than man may find

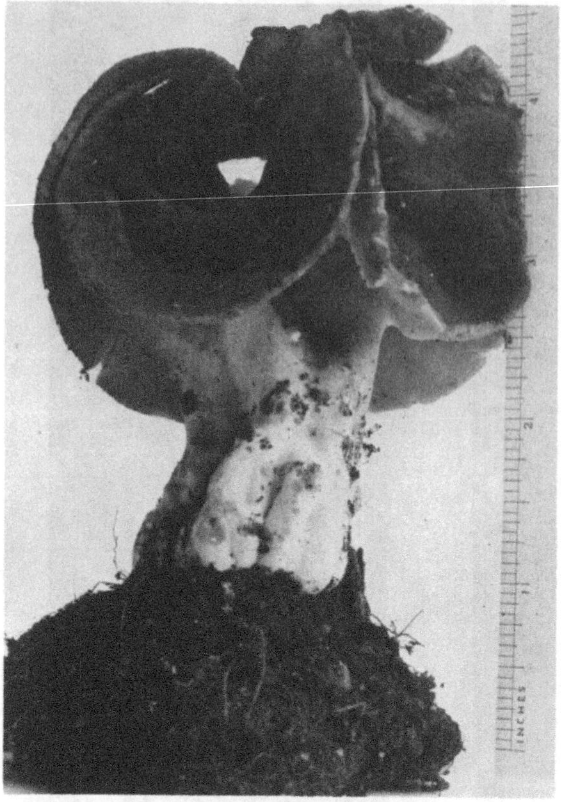

FIGURE 56. A False Morel, *Gyromitra brunnea.* This species is considered edible but the beginner is advised to leave it alone.

mushrooms a desirable food item. Most collectors soon find out, especially during the summer months, that a great variety of smaller animal life such as species of worms, insects, and slugs feed heavily on fungi. However, there are few documented instances in which fungi were deliberately fed to animals (other than experimental feedings in nutritional or toxicology studies) in quantities such that they would constitute a significant part of the animal diet.

Singer[69] stated that in Germany, during and after World War I, wild mushrooms were successfully fed to pigs, poultry, and fish, but there are no reports that this practice has been continued. This same worker suggested that the waste material from mushroom canneries (trimmed off lower portions of stipes and damaged or diseased mushrooms) could be used as fodder but cited no instance of this having been done. However, the amount of such waste material is relatively small and could make only a negligible contribution to the overall supply of domestic animal feed. Also, the expense involved in the production of mushrooms is too great to permit their use as animal feed, and at least in the U.S. the cost of labor is so high that the collection of wild mushrooms for animal feeding purposes would also be economically unsound. At this time the only places where mushrooms play a significant role in the feeding of domesticated animals are in the far north, where reindeer are an important source of meat and other commodities. Sheep are said to feed on mushrooms but the author has found no reports to this effect, and cows apparently will eat mushrooms, since Wasson[193] described his feeding of *Amanita muscaria* to several cows in Japan. Whether or not the cows hallucinated properly or had a bad trip was not noted.

Singer,[166] in his discussion of the fungus flora of the Arctic, recorded that in arctic and subarctic areas the reindeer feed on mushrooms, even digging frozen ones from under the snow, and that the local and seasonable abundance of mushrooms is one factor in choosing feeding grounds in Lapland and in the northern part of the U.S.S.R. Rautavaara[148] reported from Finland that in a good mushroom year reindeer weigh on the average 20% higher than normal while in a year when mushrooms are scarce they weigh 12% less. According to Rautavaara, about 20% of the meat production can be attributed to the feeding of fodder containing mushrooms, and in Finnish Lapland alone, this meat is valued at ten million Finnish marks annually. Apparently reindeer do not just tolerate mushrooms or eat them merely because of necessity, since Wasson[193] stated that reindeer have two obsessions – *Amanita muscaria* and human urine.

It is becoming more and more apparent that fungi, and especially fleshy fungi, may play a very important role in the feeding of wildlife. Thus, Krieger[110] was of the opinion that except for discriminating humans, red squirrels are the prime mushroom eaters, a statement borne out by the reports of Buller,[30] Cram[47] and Hatt.[87] Krieger noted that red squirrels seem to prefer the boleti but that other mushrooms are also eaten – some being stored in the forks of trees for future use. Baxter (unpublished) noted that when squirrels were offered a variety of mushrooms that he had collected, they took only basidiocarps of *Boletus edulis*, storing them in trees. According to Metcalf,[127] even the poisonous *Amanita muscaria* is eaten. The reports of Miller and Halls[128] and others indicate that gray squirrels also feed on mushrooms and I have observed fox squirrels eating *Coprinus atramentarius*. Miller and Halls stated that the western gray squirrel feeds heavily on subterranean fungi such as false truffles, but that the eastern gray squirrel favors aboveground mushrooms. Nixon et al.[136] reported that mushrooms ranked high in the gray squirrel diet of southeast Ohio, and Dudderar[57] found that fleshy fungi constituted 26% of the autumn diet and 16% of the winter diet of gray squirrels in Virginia. From these reports it is evident that wild mushrooms play a very significant role in the feeding of squirrels.

Wild mushrooms also seem to constitute an important part of the diet of deer in the U. S., which might be expected in view of the fact that they are important items in the diet of reindeer. Working in South Carolina, Urbston (c.f. Miller and Halls[128]), in his analysis of the contents of 100 deer stomachs, found that 95% contained mushrooms or other fungi and that the volume of stomach contents due to fungi varied from a trace up to 83%. Similarly, Strode[174] listed mushrooms as the most important food item for deer in the Ocala National Forest (Florida) – of the deer stomachs examined, 79% contained mushrooms, percent by volume being 24. Korschgen,[109] in his survey of the foods used by game and predatory animals in Missouri, reported that 50 out of 270 deer stomachs examined contained fungi, and Cushwa et al. (c.f. Miller and Halls[128]), who collected deer stomach samples from the major physiographic provinces of the Southeast, reported that the percent of stomachs containing mushrooms ranges from 10 to 100, depending upon the province and the season. These latter workers noted that the use of mushrooms was generally highest in the summer. This higher usage in the summer might simply be due to the greater incidence of mushrooms during the summer, but it might also be indicative of an actual preference for this type of food.

Miller and Halls[128] reported that rodents other than squirrels that eat mushrooms are white-footed and deer mice, and that such game birds as grouse, wild turkeys, and quail as well as smaller birds also feed on mushrooms and bits of mycelium. In all probability other small animals also feed on mushrooms as season and food conditions permit or necessitate. These workers have prepared a list of fleshy fungi (mostly white-spored agarics) which are eaten by one or more species of wildlife. They noted that four species of mushrooms are preferred more than others by wildlife: (1) *Lactarius piperatus*, the Peppery Lactarius, (2) *Russula roseipes*, the little Red Russula, (3) *Armillaria mellea*, the Honey Mushroom, and (4) *Suillus granulatus*, the Granulated Bolete. They issue the cautionary note that many species of mushrooms that are poisonous to man are apparently harmless to wildlife and that animal use therefore does not indicate edibility for humans. Their list, in which species are grouped according to families, is as follows:

Agaricaceae
Amanita brunnescens, A. caesarea, A. chlorinosma, A. fulva, A. muscaria, A. solitaria, A. verna
Armillaria caligata, A. mellea, A. ventricosa
Lepiota molybdites, L. naucina, L. procera
Russula borealis, R. crustosa, R. delica, R. foetans, R. roseipes, R. variata, R. virescens
Lactarius indigo, L. piperatus, L. volemus
Cantherellus cibarius, C. floccosus
Hygrophorus coccineus, H. russula, H. tennesseensis
Clitocybe cartilaginea, C. illudens, C. parilis
Collybia sp.
Laccaria laccata, L. ochropurpurea
Tricholoma flavobrunneum, T. flavovirens, T. personatum, T. resplendens
Agaricus campestris, A. placomyces, A. sylvicola
Cortinarius distans, C. evernius, C. squamulosus
Pluteus cervinus
Clitopilus prunulus

Boletaceae
Boletus betula, B. rubelus
Boletinus pictus
Gyroporus cyanescens
Pulverboletus retipes
Strobilomyces floccopus
Suillus granulatus, S. luteus

Lycoperdaceae
Calvatia gigantea

Clavariaceae
Clavaria sp.

Hymenogastraceae
Rhizopogon sp.

Helvellaceae
Morchella esculenta

Miller and Halls stated that wildlife apparently prefer agarics which are classified in the white, pink, yellow-brown, and purple-brown spored groups and that they had not observed any use of black-spored species by wildlife. This failure to use black-spored species was attributed to the fact that as some of the mushrooms of this group (*Coprinus* spp.) mature, the pilei and gills liquify through a process of autodigestion. They apparently were not in the right place at the right time, because I have observed the fox squirrel in northern Illinois eating the pilei (but not the stipe) of the Inky Cap (*Coprinus atramentarius*) with apparent relish. The fact that the pilei were beginning to deliquesce did not seem to make any difference. It is of interest to note that of the 58 fleshy fungi reported by Miller and Halls to be eaten by wildlife, 6 species have been reported by at least one author to be poisonous to man (see Table 2 in the section on Mushroom Poisoning and Poisonous Mushrooms which follows). This adds substance to the view that fungi which are poisonous to carnivorous animals may often be eaten with impunity by herbivorous animals.

The above listing of species of fleshy fungi which may be eaten by wildlife makes a good starting point and should prove of some use to individuals involved in wildlife management, but it is highly unlikely that it represents anything approaching a complete listing of species that wildlife may use for food. As further information accumulates it may well develop that such fungi play a much more important role in a variety of natural food chains than is now realized. However, by its very nature, the amassing of accurate and complete information of this type is a very slow process, since it depends upon the interested individual being at the proper place at the proper time and in the proper season; hence it may be many years before complete information is at hand. In the interim, since there is presently so much concern being expressed about the changes that man has made and is making in the eco-system, it might be the better part of good judgment to give some thought to the possible consequences if widespread major changes are made in fleshy fungus habitats with resultant changes in mushroom flora.

MUSHROOM POISONING AND POISONOUS MUSHROOMS

There is probably no aspect of mycology in which there is such a great need for extensive research pointed toward the objective of clarifying conflicting reports as in the area which is concerned with mushroom poisoning. In spite of the fact that mushrooms have been a part of the human diet for a great many centuries, one can still easily demonstrate that there are two rather strongly contrasting points of view: the first holds that the number of poisonous mushrooms is no more than a dozen or two, the second is that agarics are primarily a group of poisonous plants. The truth is probably somewhere between these two extremes, although on the basis of information accumulated to date, it would seem that the second point of view is an unrealistic and uninformed one and that, in fact, the non-poisonous species far out-number the poisonous ones. There are a vast number of published reports of mushroom poisoning which have certainly not been verified, many of them by physicians who could scarcely be expected to identify accurately as to species a type of organism with which they are largely unfamiliar, others by mycologists with no medical training. Often the physician involved in a suspected case of mushroom poisoning is furnished only with a questionable oral description of the suspected mushroom, a piece of a mushroom or often an old, rather badly mauled specimen.

It is probable also that there are reported instances of mushroom poisoning which have been attributed to a correctly identified species, but which may have actually been due to the ingestion of over-mature specimens which had been spoiled by the growth of bacteria or mold-type fungi. The eating of old or spoiled mushrooms (or any other type of food for that matter) could unquestionably lead to stomach upset and/or intestinal disturbances, but this certainly does not indicate that the mushroom *per se* is poisonous. Still another factor could also be the cause of some confusion. Different individuals react differently to the foods they ingest and there is often a considerable amount of truth in the old adage that "one man's meat is another man's poison." Some people are allergic to chicken and cannot eat even a small amount without becoming ill, but most are not. The same thing applies to strawberries, grapes, oysters, and probably every item of food that man eats if all of the facts were known. There is no reason to believe that such individual differences do not also apply to mushrooms (Dearness,[52] Simons[165]). Such a point of view is borne out by the fact that some people deliberately seek out and collect the False Morel, *Gyromitra esculenta*, which they consider a harmless delicacy and eat with impunity, while others may be fatally poisoned by this species. This same anomalous situation prevails with respect to the green-spored Morgan's Lepiota, *Lepiota morgani*.

Fischer[68] attributed the many cases of mushroom poisoning which occurred in the U.S. early in this century to the fact that there had been a great influx of immigrants from southern Europe, who, accustomed to gathering and eating the wild species they had known in Europe, sometimes confused poisonous American species with the edible ones they had known in Europe. Thus, Peck[141] reported that of the more than 30 mushroom poisoning deaths that occurred in and near New York City in 1911, nearly all of them involved people who had such origins. Apparently this applied only to people who originated in southern and eastern Europe, since people of Anglo-Saxon origin are usually considered to be "mycophobes," and Wasson[193] used this same appellation in connection with peoples of Germanic or Celtic origin. However, instances of mushroom poisoning were by no means restricted to peasant immigrants from Europe, because prior to the time of Peck and Fischer, poisonous mushrooms were reported to have claimed such notable victims as Pope Clement VII, Emperor Jovian, Emperor Claudius, King Charles VI of France, Czar Alexis of Russia (or his widow), and the entire family of the early greek poet Euripedes (Ford and Clark,[73] Fischer,[68] Ford,[71,72]). What with the sometimes peculiar antics and known intrigues of many high-placed Romans of imperial days and the jockeying for position among some of the early Popes, one has a right to question whether the Roman and Papal deaths were truly accidental or had been planned – a question which might also justifiably be raised in connection with the poisoning death of a member of the Russian royal family.

The practice of collecting and eating wild mushrooms is unquestionably far more widespread in Europe than it is in the U. S. Not only do many Europeans collect edible wild mushrooms for their own use, but also many are collected so that they may be offered for sale to others. Thus, Duggar[58] reported that in 1901 Munich was the principal mushroom market of Europe and that 31 different species were sold there commercially. Apparently a certain measure of control was exercized over the species of mushrooms which could be offered for sale in public markets. The practice of commercializing wild mushrooms continues today in Europe, and in gourmet shops in the U. S. it is possible to purchase canned morels imported from France, pickled or marinated Select Boletus from Poland, canned Chantrelle from Sweden, etc. Much of the dried mushroom flavoring offered for sale in England is prepared from dried *Boletus edulis* (Select Boletus) collected on the continent.

According to Fischer,[68] quantitative knowledge of deaths caused by mushroom poisoning begins with Paulet, who reported that during the years 1749 to 1788 there were 100 such deaths in the Paris vicinity alone. Fischer then cited reports of various other workers which indicated that there had been nearly 1,500 recorded cases of mushroom poisoning by early in the twentieth century from this and other parts of the world. In all probability the actual number of cases was greater than 1,500 because this was still during the period that man was finding out by a trial and error method which mushrooms he should eat and which ones he should not, and it is probably highly improbable that all such deaths were reported. It should also be recognized that some of these earlier reports were undoubtedly erroneous because many of them were made at a time when modern, safe food preservation methods were not known, and many deaths that were reported as cases of mushroom poisoning could possibly have been due to the ingestion of food (mushrooms and otherwise) which was spoiled and thus could not justifiably be ascribed to poisonous mushrooms.

In addition to reports of mushroom poisoning from Europe and North America, such cases have been reported from every other continent. Thus, Cleland[41] and Trotter[178] have reported such occurrences in Australia, Miura[130] in Japan, Seal et al.[161] in India, Mendoza and Leus-Palo[126] in the Philippines, Cullinan and Henry[50] in Africa and Vicuna et al.[186] and Munilla et al.[132] in South America.

Although it is known that prior instances of mushroom poisoning had occurred (Jenkins[102]), the first report of mushroom poisoning to appear in the medical literature of the U. S. was that of Cheney.[35] In 1910 Murrill[133] estimated that the annual deaths from mushroom poisoning in the U. S. at 50 or more, and several years later Fischer[68] quoted this figure but expressed the belief that not even 10% of the cases of mushroom poisoning find their way into the medical literature. Dreisbach[55] stated that over 100 fatalities occur each year (presumably in the U. S.) from eating poisonous mushrooms, and Block et al.[23] estimated that there were 50 fatalities annually in the U. S., a figure inadvertently also cited by Emmons[61] in 1961. However, Simons[165] stated that this estimate is much too high. In 1962 Emmons[62] stated that "Although an estimate of 50 deaths per year from mushroom poisoning may have been a conservative estimate for the period just before and after World War I, it is obviously erroneous to project such an estimate into the present." Emmons then stated that in the U. S. there were 8 deaths in 1957, 4 in 1958, and 2 in 1959. It would seem that for more recent times, Simons and Emmons are correct, since Buck[29] found a total of only 35 fatalities in the American literature during the years 1924 to 1961. In spite of the fact that Anglo-Saxons are usually considered to be mycophobes, they are not uniformly so, since the incidence of deaths from mushroom poisoning seems at least as high and perhaps somewhat higher in Great Britain than in the U. S. Thus, Birch[20] found 38 fatal cases in England and Wales during the years 1920 to 1945. Similarly, Dubash and Teare[56] noted that 21 cases of *Amanita phalloides* poisoning were reported by the London daily press over the 15 year period immediately prior to 1945. While Wasson[193] considers people of Germanic origin to be mycophobes, apparently some Germans are willing to eat mushrooms, since Germany had 200 deaths from mushroom poisoning in 1948 (Lewes[114]).

Of the 35 American fatalities occurring during the years 1924 to 1961 which were cited in Buck's review, the responsible mushrooms were as follows:

Gyromitra esculenta	7
Amanita phalloides	11
Amanita pantherina	1
Psilocybe baeocystis	1
No species name recorded	15

In the light of the above statements it might seem especially foolhardy for anyone to recommend that any person eat any wild mushrooms of any type at any time. However, this would represent a very narrow and unrealistic point of view, since the known facts at this time most certainly do not bear out the concept that the majority of fleshy fungi are poisonous. This can be easily demonstrated by examining Table 1, in which are summarized numerical data based upon evaluation of reports taken from seven separate handbooks and manuals, which show that in every instance the authors have listed a far greater percentage of species of mushrooms as edible than they have listed as poisonous.

Even a casual inspection of the figures presented in Table 1 shows that in every instance the author of a specific manual has listed many more species as edible than he has listed as poisonous. It might be assumed that this is due to the fact that these manuals or handbooks are weighted specifically in favor of the presentation of descriptions of edible species, but this is not necessarily the case. Thus, if the most extensive of the manuals used in assembling the data of Table 1 is considered, it will be noted that Kauffman[106] discussed 884 species, of which he listed 20% as edible; however, he listed only 3% as being

TABLE 1

A Comparison of the Numbers of Edible and Poisonous Species of Fleshy Fungi. Based on the Reports of Zeitlmayr, Kauffman, Lange and Hora, Smith, Christensen, Atkinson, and Krieger.

	Number of species		
Author	Discussed*	Listed as edible	Listed as poisonous
Zeitlmayr	70	45	9
Kauffman	884	180	30
Lange and Hora	594	104	25
Smith	222	98	21
Christensen	309	128	6
Atkinson	208	81	8
Krieger	214	163	11

*An attempt has been to include here only those fleshy fungi which conceivably could be chewed and thus considered to be potential food items.

poisonous. In addition, he listed another 36 species (4%) as suspected or included with them some sort of cautionary note. In his excellent attempt to describe the complete mushroom flora of Michigan, Kauffman listed six times as many edible species as he listed poisonous ones. Even if the 36 species which he listed as suspected are grouped with the poisonous ones, the edible species are almost 3 times as numerous as the poisonous and suspected ones together. Further substance is added to this view by the report of Guba.[85] This latter author prepared a list of over 800 species of mushrooms which had been gathered in New England since 1896 and identified by members of the Boston Mycological Club; of these species, Guba considered only 53 to be poisonous. Based upon Heim's estimate that there are 85,000 different species of fungi in the world (which may be quite a conservative estimate – see Martin[122]), Wasson[190] has estimated that there are 30,000 species of larger fungi (mushrooms). Ford[72] listed 79 species which he considered poisonous, a very small percentage of the total if the 30,000 figure is correct.

Regardless of how the matter is approached, at first sight it would appear that statistically the chances of collecting a poisonous species are considerably less than the chances of finding an edible one. However, such reasoning could be dangerous since it completely disregards any possible differences that might occur between species with regard to the numbers of individuals produced. Thus, it is rare that more than a few individual basidiocarps of the edible *Volvaria bombycina* (Silky Volvaria) are found, whereas in season in the proper habitat literally bushels of the poisonous *Amanita muscaria* often may be collected in a relatively small area. By way of example, the experience of Puharich[145] may be cited. This author, who apparently had seen *A. muscaria* only once before, reported collecting 235 fresh specimens of this species in one day in western Massachusetts. Hence, in the matter of the probability of finding edible vs. the finding of poisonous mushrooms, reliance upon what appear to be very favorable statistics could easily result in becoming a statistic oneself, in the obituary column! Or, more humiliating still, providing the excuse for sentimental survivors to erect a tombstone recording another mushroom fatality (Jenkins[102]). Cann and Verhulst[32] reported that accidental ingestion of unidentified mushrooms accounts for less than 2% of the cases reported to the National Clearing House for Poison Control Centers.

As pointed out in the section on the Collection and Identification of Edible Mushrooms, no one should ever eat any mushroom until they are absolutely sure as to its identity. In fact, it was suggested that for the person interested in collecting and eating wild mushrooms, perhaps it would be advisable to first learn to recognize the poisonous species. Table 2 has been prepared to assist in this. In this table, each species of mushroom reported by each of the seven authors listed in Table 1 is recorded. In addition, the poisonous species listed by Fischer[68] are also included. The scientific names used in the table are those listed by the separate authors, a practice which makes the list of poisonous species appear to be somewhat longer than it actually is. For example, five of the eight authors reported *Gyromitra esculenta* as a poisonous species, while one reported *Helvella esculenta* as poisonous; *G. esculenta* and *H. esculenta* may well be different names for the same species. Similarly four authors reported *Lepiota morgani* as a poisonous species, but Smith[171] does not list this species. Instead, he reported *Lepiota molybdites* as poisonous and placed *L. morgani* into the synonomy of this species.

From Table 2 it may be seen that 76 species of fleshy fungi have here been recorded as poisonous. Because of different names being applied to the same fungus in at least two instances, this number should probably be reduced to 74. Further evidence in favor of the concept that poisonous mushrooms are in the minority may be drawn from this table. Of the 76 fleshy fungi listed, 46 are cited by only 1 author out of 8. Of the 23 genera listed as containing one or more poisonous species, only 19 are genera of agarics, a rather small harvest considering the fact that Kauffman[106] described 51 different genera of agarics in his book.

The situation observed for genera may also be observed at the specific level; 76 different species are listed in Table 2 but of these only 69 are species of agarics, again, a rather skimpy harvest of poisonous species in the light of the 884 species of agarics that were discussed by Kauffman. Aside from the genus *Amanita,* which contains a high proportion of poisonous species, in most genera which contain poisonous species, these latter are in

TABLE 2

Reports of Poisonous Mushroom Species. Based upon the Listings of Atkinson,[9] Christensen,[37] Fischer,[68] Kauffman,[106] Krieger,[110] Lange and Hora,[112] Smith,[171] and Zeitlmayr.[204]

		Reference Number							
	Spore Color	106	68	171	9	37	110	204	112
Amanita bisporigera	W	+							
Amanita chlorinosoma	W		+						
Amanita chrysoblema	W	+							
Amanita cothurnata	W			+	+?				
Amanita frostiana	W				+				
Amanita mappa	W	+				+			
Amanita muscaria	W	+	+	+	+	+	+	+	+
Amanita pantherina	W		+	+			+	+	+
Amanita phalloides	W	+	+		+	+	+	+	+
Amanita spissa	W	+							
Amanita spreta	W	+	+	+					
Amanita verna	W	+	+	+	+	+	+		
Amanita virosa	W	+	+		+		+	+	+
Amanitopsis volvata	W	+					+		
Boletus luridus			+						
Boletus satanus			+						+
Cantharellus floccosus	W			+					
Clavaria formosa									+
Clavaria pallida								+	
Clitocybe dealbata	W								+
Clitocybe illudens	W	+	+	+	+	+			
Clitocybe morbifera	W	+	+						
Clitocybe nebularis	W		+						
Clitocybe rivulosa	W								+
Clitocybe sudorifica	W		+						
Entoloma lividum	P							+	
Entoloma sinuatum	P						+		+
Gyromitra esculenta			+		+?		+	+	+
Hebeloma crustuliniforme	YB			+					
Helvella esculenta				+					
Helvella infula				+		+			
Hygrophorus conicus	W			+			+		
Inocybe asterospora	YB								+
Inocybe fastigiata	YB								+
Inocybe fibrosa	YB	+				+			
Inocybe geophylla	YB								+
Inocybe griseo-lilacina	YB								+
Inocybe infelix	YB	+	+				+		
Inocybe infida	YB		+						
Inocybe jurana	YB								+?

TABLE 2 (continued)

Reference Number	Spore Color	106	68	171	9	37	110	204	112
Inocybe lacera	YB								+
Inocybe langei	YB								+?
Inocybe maculata	YB								+
Inocybe napipes	YB								+
Inocybe patouillardii	YB							+	+
Inocybe praetervisa	YB								+
Inocybe radiata	YB	+							
Inocybe squamosa	YB								+?
Inocybe sp.	YB			+					
Lactarius chrysorheus	W	+				+			
Lactarius cilicioides	W	+							
Lactarius lignyotus	W	+							
Lactarius pyrogalus	W	+							
Lactarius rufus	W	+							
Lactarius scrobiculatus	W	+							
Lactarius torminosus	W	+	+						
Lactarius trivialis	W			+					
Lactarius uvidus	W	+				+			
Lactarius vellereus	W			+					
Lepiota clypeolaria	W			+					
Lepiota molybdites	W			+					
Lepiota morgani	W*	+	+		+?	+			
Marasmius peronata	W	+							
Marasmius urens	W	+							
Naematoloma fasciculare	PB			+					
Panaeolus campanulatus	B			+					
Panus stipticus	W	+							
Pholiota autumnalis	YB						+		
Russula emetica	W		+		+				+
Russula foetans	W			+					
Stropharia coronilla	PB			+					
Stropharia hornmanii	PB								+
Tricholoma vaccinum	W			+					
Tricholoma venenata	W	+							
Volvaria gloiocephala	P	+							
Volvaria speciosa	P	+							

*Green-spored but classed with white-spored group
? = suspected

the minority. Thus, Kauffman discussed 41 species of *Lactarius,* 50 species of *Russula,* and 42 species of *Clitocybe,* but in Table 2 only 10, 2, and 6 species, respectively, of these 3 genera are listed as poisonous.

In the first column following the species names in Table 2, the spore color of each listed species is indicated by letter. Of the 69 species of agarics listed, the distribution of poisonous species with regard to spore color is as follows:

white-spored	42
purple-brown-spored	3
pink-spored	4
black-spored	1
yellow-brown-spored	19

From the above listing it is quite evident that the great majority of species of poisonous agarics have white spores (see *A* in Figure 9). Not only that, but some of the most poisonous of all fungi (the deadly Amanitas) are in the white-spored group. The Fly Agaric (*Amanita muscaria*) is the only species upon which there is complete agreement between all eight authors concerning whether or not it is poisonous. This is most curious since today there are individuals who stoutly maintain that *A. muscaria* is not poisonous. Smith[171] does not list *Amanita phalloides* as a poisonous species, but this is obviously not because he does not believe it to be poisonous but because he was concerned only with American species and at that time it had not been definitely established that *A. phalloides* (which is primarily a European species) does occur in the U. S. It has now been established (e. g., Smith, personal correspondence) that this deadly European species does occur on both the east and west coasts of the U. S.

The lack of agreement among authors in Table 2 is rather more apparent than real, because in some instances the failure of one author to report a species as being poisonous may simply result from the fact that he did not treat that particular species in his manual. There also exists the possibility that some of the species listed (and especially those species cited by only one author) may not be poisonous and the report may be erroneous due to a misidentification or because someone became ill from eating a mushroom that had been spoiled by other microorganisms. Also, in view of the widespread mycophobia among Anglo-Saxons, some cases of "mushroom poisoning" undoubtedly arose in the mind rather than in the stomach. None the less it would be the better part of good judgment for the beginning mycophagist not to eat any mushroom which any authority has listed as being poisonous.

There are a number of species about which there are still justifiable differences of opinion regarding whether or not they are poisonous. One such species is *Lepiota morgani* (the *Lepiota Molybdites* listed by Smith[171]). This species, the only green-spored mushroom (but classed in the white-spored group), is regularly eaten without any ill results by some people. On the other hand, still other people are poisoned by *L. morgani.* It is now quite obvious that different individuals react differently to the ingestion of this mushroom. For that reason the individual who has had no previous personal experience with this species should probably leave it alone. On the other hand, it is a quite large mushroom and in some areas very abundant in season and so he might wish to test his reactions to it in the manner used by Smith, the late G. W. Martin, and other mycologists. This method is to eat a very small amount on the first encounter and see if anything untoward happens. If there are no disagreeable results, a larger amount may be eaten on the second encounter. In this manner a gradually larger amount may be ingested and it can be established whether or not that particular individual can safely eat *Lepiota morgani* without becoming ill. Historically it has been in this general manner that the information regarding the edibility of a great many fleshy fungi has been accumulated (sometimes with unhappy results) over the centuries. However, it should be borne in mind that information regarding poisonous fungi also has been accumulated in the same manner (intentionally or not) and that the man who first proved that *Amanita phalloides* was poisonous probably did not live to boast about his new knowledge.

On the basis of published evidence to date, it is apparent that the most poisonous species of mushrooms are members of the genus *Amanita.* Of these species, *Amanita phalloides* and some of its close relatives such as *A. verna, A. bisporigera,* and *A. virosa* are the deadliest. Fischer[68] stated that *A. phalloides* and related species are responsible for at least nine tenths of all fatal cases of mushroom poisoning. Rolfe and Rolfe[152] also ascribe 90% of fatal cases of mushroom poisoning to these species, and Wieland[197] stated that more than 95% of the cases of fatal mushroom poisoning are due to *A. phalloides* and its white relatives. VanderVeer and Farley[183] concurred in this since they stated that nearly all severe cases of mushroom poisoning in the U. S. occur as a result of the ingestion of some species of *Amanita*.

Ramsbottom[146] set the mortality rate of *A. phalloides* poisoning as high as 90%, but Henrici[94] reported it as being 50 to 70%. The extreme toxicity of this species is evidenced by Bessey's[19] estimate that as little as a one cubic centimeter-sized piece will cause severe illness or even death.

Types of Mushroom Poisoning

In 1912 Frey (cf. Fischer[68]) commented that the whole study of mushroom toxicity was still very much in the dark and that it was on the same plane as it had been 30 years ago. Apparently Frey was unaware of the attempt of Ford[70] to divide mushroom poisoning into three general types based upon the characteristics of the symptoms caused by the poisons contained in these fungi. Later, Ford[71,72] classified mushroom poisoning (mycetismus) into five categories rather than the original three, and Grossman and Malbin,[82] Gray,[79] and Emmons[61] among others accepted these categories, although there is now some question as to whether or not these may need reevaluation and possible revision. Wasson[190] objected to the use of the word mycetism and justifiably raised the question as to why a term that should properly mean the quality of being a fungus should be attached solely to the toxic properties of a few mushrooms; he termed such usage as "mycophobia" at its most extreme. Perhaps others are arriving at this same point of view, since in his *Handbook of Poisoning: Diagnosis & Treatment*, Dreisbach[55] does not use the word mycetismus. The five categories of mycetismus proposed by Ford are

1. **Mycetismus gastrointestinalis** – The symptoms of this type of mushroom poisoning are nausea, vomiting, and diarrhea – much like various types of mild, non-fatal food poisoning. The severity of the symptoms may vary greatly, probably primarily with the individual but also with the quantity of mushrooms ingested. In all probability (although it is a virtually unexplored point) severity of poisoning might vary from strain to strain of the same species of mushroom. Morphologically indistinguishable strains of the same species of mold are notoriously different in their biochemical capabilities. Thus, one strain of the ubiquitous *Aspergillus niger* may synthesize citric acid in such large quantities that a commercial fermentation process is feasible, while another strain of the same species may produce such small amounts of this acid that it is difficult to detect among the metabolic products. Similarly, differences may be demonstrated between different isolates of the same species of higher Basidiomycetes. Thus, Gray[78] demonstrated that different isolates of *Polyporus schweinitzii* differ markedly in their capabilities to decay wood, and currently in my laboratory there are several isolates of the edible Oyster Mushroom (*Pleurotus ostreatus*) which vary in their abilities to use cellulose as the sole source of carbon. Therefore, there is no reason to believe that different strains of the same species of poisonous mushroom would not vary with respect to the amount of toxic material which they synthesize. Environmental conditions under which the mushrooms developed might also be expected to greatly influence such syntheses.

The symptoms of mycetismus gastrointestinalis terminate rapidly and usually spontaneously, and the patient is commonly restored to health in a day or two. According to Ford there are practically no fatalities, and the species of fungi which are usually responsible are *Russula emetica, Boletus satanus, B. minato-olivaceus, Lactarius torminosus, Entoloma lividum* and *Lepiota morgani* (Figure 57). In all probability, a number of other mushroom species should be added to this list because there are various reports which ascribe mild gastrointestinal disturbances to the ingestion of species other than those listed by Ford. However, with the state of our knowledge as it is now, it would be best to leave the list as it now stands until further, more exact information is at hand. Also, it seems possible that some species of mushrooms have been implicated as the cause of illness when it may have been simple gluttony or a gastronomic enormity, not the mushroom *per se* which caused the illness. It can be demonstrated that there are toxins in many of the common foods eaten by man, especially the legumes (Liener,[115]). When such foods are eaten in reasonable amounts there are no problems because the toxins are present in such low concentration; however, it is conceivable that in cases of over-eating, the discomfort which follows may be due in part to the presence of greater concentrations of a toxin. The most extreme and impractical example of this type is the report of Robinson and Davidson[151] in which it is stated that morel mycelium is safe to eat in amounts up to 60 pounds per day.

There is a strong possibility that what Ford called mycetismus gastrointestinalis is not poisoning in the strict sense but on the contrary is merely a food allergy – certain individuals being sensitive to different types of mushrooms. There are many types of mushrooms and the allergic reactions of different individuals vary over a tremendously wide range. Therefore, it would not be at all surprising to find that certain people are allergic to certain species of mushrooms. Thus, I eat the Velvet-Stemmed Mushroom, *Collybia*

FIGURE 57. Morgan's Lepiota, *Lepiota morgani.* This species is poisonous to some individuals but since it is green-spored it is easily identified. (Photograph by A. H. Smith.)

(Flammulina) velutipes, without ever encountering any problems but I have a friend who becomes ill if she eats this particular species. Similarly, Simons[165] reported that he becomes very uncoordinated if he eats the Morel relative, *Verpa (Ptychoverpa) bohemica*. Mullendore[131] must have been aware of similar individual reactions to *V. bohemica* since he listed this species as "edible, but be cautious." None the less this reaction does not appear to be general since some individuals seek *V. bohemica* as a desirable edible species. Simons[165] also reported that one of his acquaintances becomes quite ill (mycetismus gastrointestinalis?) if he eats only a small amount of the Common Cultivated Mushroom (*Agaricus bisporus*) – a species eaten with relish and in safety by millions of people. Bergoz and Righetti[18] have described an unusual case of selective intolerance to mushrooms, the symptoms of which could be reproduced by the ingestion of trehalose. They suggested that this intolerance was due to a deficiency of intestinal trehalose, but could not determine the level of intestinal disaccharidase activity.

2. **Mycetismus choleriformis** – More has probably been written about this type of mushroom poisoning than any other type (with the possible exception of *Amanita muscaria* poisoning, concerning which there is still some difference of opinion), although it is generally not referred to under the name applied to it by Ford. The most unfortunate aspect of this often fatal type of poisoning is that the symptoms do not appear until a relatively long time after ingestion, at which time it may be too late to do much for the victim except to help make his passage from the world of the living as painless as possible. During the early stages gastrointestinal disturbances are commonly predominant, usually appearing 10 to 15 hours after the mushrooms were ingested. The patient has violent cramps which are accompanied by purging and vomiting. He then rapidly loses strength and weight and may die in two to five days. Degenerative lesions of heart muscle, liver, and kidneys occur. Poisoning of this type is usually caused by *Amanita phalloides* in Europe, although in the U. S. its close relatives (Figures 58, 59, 60 and 61), especially the white forms such as *A. verna* and *A. virosa*, are more commonly involved, but Ford[72] also included *Galerina (Pholiota) autumnalis* and *Hygrophorus conicus* here. Peck[141] reported a case in which a mother and two children ate *Galerina autumnalis*; all three became ill and the two children died. Ford and Sherrick[74] found that *G. autumnalis* was poisonous to guinea pigs and

FIGURE 58. *Amanita bisporigera*, one of the deadly white Amanitas – a close relative of *A. phalloides*. (Photograph by A. H. Smith.)

FIGURE 59. The Destroying Angel, *Amanita virosa*. (From Groves.[83] By permission of the Canada Department of Agriculture.)

rabbits, and their postmortem findings resembled those caused by *A. phalloides*. No further cases of *Galerina* poisoning were reported until Grossman and Malbin[82] reported poisoning due to *Galerina venenata* and also noted the similarity of the poisoning to that caused by *A. phalloides*. Tyler et al.[182] demonstrated the presence of amatoxins (see below) in these species, and Benedict et al.[16] and Benedict and Brady[13,14] found that the mycelia of *Galerina marginata* can synthesize amanitins when cultured in synthetic medium. From these reports it now seems apparent that there are biochemical similarities between the deadly Amanitas and certain species of *Galerina*.

The early mycologists and medical men in the U. S. generally referred all species of deadly Amanitas in this country to *Amanita phalloides*, which was apparently considered to consist of a variety of different forms. Thus, Murrill,[134] who attempted without success to change the name of this species to *Venenarius phalloides*, described this species as varying in color from white to yellow, yellowish-green, green, gray, brown, or blackish. The various color forms were subsequently classified into a number of different species and for some years there was doubt as to whether or not the true *Amanita phalloides*, which is a predominantly European species, occurs in the U. S. at all. However, it was reported that the species does occur rarely in California and it was later found in the Pacific Northwest; it is now known from both coasts. Tyler et al.[181] consider the Pacific Northwest specimens to be morphologically identical with the European form but chemically somewhat different. It would now appear that the true *A. phalloides* does occur in the U. S., but since related species (especially the white ones) seem far more abundant and hence more likely to be encountered, they are the species which have been depicted in the accompanying pictures.

For many years if an individual was poisoned by ingesting *Amanita phalloides* or one of its deadly relatives, there was little that could be done for him except to make him as comfortable and free of pain as possible and hope that he might be

FIGURE 60. The Spring Amanita, *Amanita verna.* (Photograph by G. F. Atkinson.)

FIGURE 61. *Amanita brunnescens,* the American brown counterpart of *A. phalloides.* Note scarcity of scales on the pileus at left. (Photograph by A. H. Smith.)

one of the lucky minority that escapes death after eating these mushrooms. If there is any doubt regarding the horrible suffering to which the patient is subjected in this type of poisoning, such accounts as those of Fischer,[68] Simons,[165] and others should be consulted.

According to Ford and Clark[73] and Fischer,[68] the first attempt to isolate the toxic principle of *Amanita phalloides* was probably that of Letellier in 1826. This early worker obtained a heat-stable substance which was termed amanitin. Various other attempts have been made over the years, and the chemical structures of the toxic substances obtained from the *A. phalloides* group of mushrooms have been determined (see Wieland and Wieland[198] and Wieland[197]). These toxins are of two general types: the phallotoxins which are cyclic heptapeptides and the amatoxins which are cyclic octapeptides (Figure 62). Phalloidin is the most abundant phallotoxin and α-amanitin and β-amanitin are the most important amatoxins – the former being neutral, the latter acidic. Phalloidin has a much lower toxicity than the amatoxins, the lethal dose of the latter for man being estimated as about 0.1 mg/kg body weight. Therefore, the amatoxins are primarily involved in the poisoning of man, and the amatoxin contents of various species of *Amanita* have been determined (Block et al.,[23,24] Benedict et al.,[15] Tyler et al.,[181] Wieland[197]).

Apparently the amanitins can attack many organs, but the liver and kidneys are usually most seriously involved. When a minimum lethal dose of α-amanitin is injected into a mouse, the proximal convoluted tubules of the kidney are damaged but there is no effect on the liver. However, when the minimum lethal dose is exceeded, attack on the liver occurs (Fiume and Wieland[69]). Myler et al.[135] reported similar proximal convoluted tubule damage in man in a case of human poisoning. Abul-Haj et al.[1] reported on a fatal case of poisoning in an 8-year-old boy and were able to identify phalloidin and phalloin from liver extracts by ultraviolet-absorption spectral characteristics

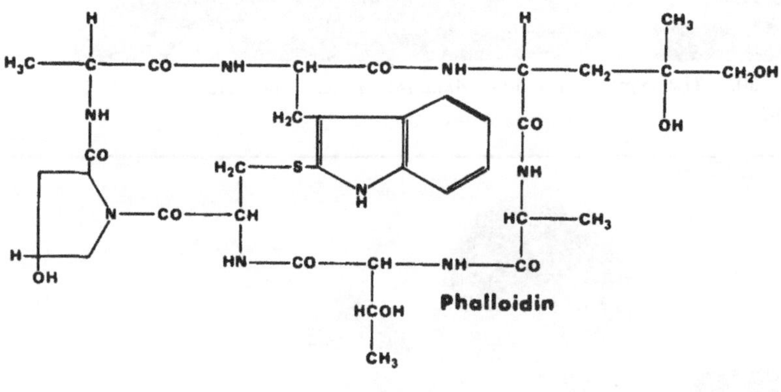

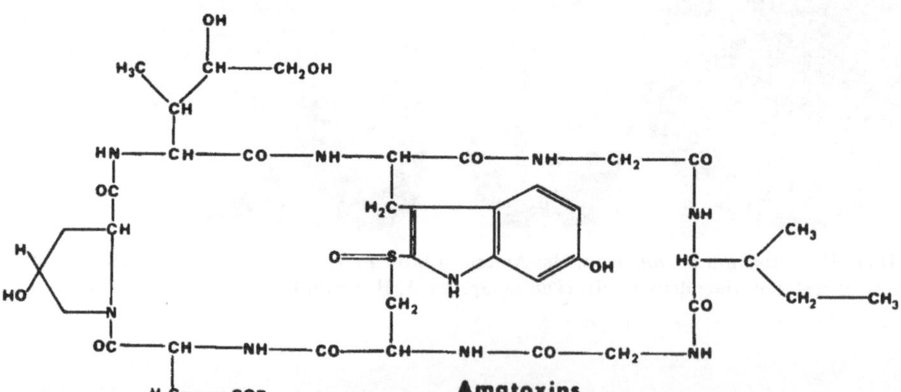

FIGURE 62. Structural formulae of phalloidin, the most abundant of the phallotoxins, and the amatoxins. In α-amanitin R = $-NH_2$; in β-amanitin R = -OH.

and by chromatographic separation; chromatography also indicated that β-amanitin and possibly α- and γ-amanitin were present. These workers also performed histologic examinations and reported that they had the following characteristics: marked gastrointestinal edema and congestion; hemorrhagic gastroenteritis; hyperplasia of intestinal lymphoid tissue and deep abdominal lymph nodes; focal hemorrhages and acute inflammation of the liver; cloudy swelling and fatty degeneration of the heart and kidneys; and toxic damage to blood vessels. They noted that a striking and consistent finding is fatty degeneration and necrosis of the liver, and that the brain shows edema, congestion, punctate hemorrhages, perivascular lymphocytic infiltration, and a degenerative process affecting nerve cells and glia.

There is still some question as to how *Amanita phalloides* poisoning should be treated or even if any treatment is really effective. An antitoxin was prepared at the Pasteur Institute (Dujarric de la Rivière[59,60]), but Grossman and Malbin[82] stated that there is no clear record of its effectiveness in man. Limousin and Petit[116] found that cats died when fed *A. phalloides* but that rabbits did not. They then theorized that secretions from the rabbit's stomach counteracted the effects of the poison and found that cats could survive the ingestion of lethal doses if they were accompanied by fresh rabbit stomach and brain. Fresh rabbit brain and stomach were then minced together and fed to patients suffering from *A. phalloides* poisoning, and Limousin and Petit considered them to be improved by such therapy. However, a subsequent report (Melandro[125]) on this method of treatment is not encouraging. Dreisbach[55] recommends (1) removal of ingested mushrooms by gastric lavage or emesis followed by catharsis (at least three times the first day and then daily until symptoms begin to subside), (2) administration of atropine subcutaneously (2 mg every hour); however, if symptoms of atropine poisoning predominate it should be administered cautiously and discontinued if symptoms do not abate, (3) careful control of fluid and electrolyte balance, (4) administration of large amounts of dextrose solution to protect liver from further damage, (5) pain control by morphine sulfate (10 mg subcutaneously every six hours as necessary); codeine for younger children, (6) treatment of aneuria, (7) use of succinylcholine or chlorpromazine to control convulsions, (8) reducing fever by active cooling.

More recently it would appear that a more sophisticated and effective method of treating *Amanita phalloides* poisoning may be developed. There is some evidence that thioctic acid (α-lipoic acid) may be effective in the treatment of this type of poisoning. Thus, Kubicka[111] reported the saving of 39 out of 40 patients through the administration of 300 mg per day of thioctic acid until serum glutamic-oxalacetic transaminase (SGOT) and serum glutamic-pyruvic transaminase (SGPT) levels decreased. Dabski et al.[51] reported that thioctic acid treatment was ineffective in two cases but in this instance the treatment was started late and the dose was only 200 mg per day. In suspected cases of *A. phalloides* (or its close relatives) poisoning, SGOT and SGPT levels should be monitored in order to determine early if liver damage has occurred. According to Schmidt et al.,[156] thioctic acid is an essential cofactor in several enzymatic processes and the LD_{50} for mice is 160 to 275 mg/kg. Simons[165] has pointed out that thioctic acid does not have FDA approval for treatment of any disease.

3. Mycetismus nervosus – In this type of so-called mushroom poisoning symptoms typically appear much sooner after ingestion than in mycetismus choleriformis – usually in two or three hours or less. There are violent gastrointestinal symptoms accompanied by contracted pupils, profuse perspiration and salivation, convulsive movements of muscles, and often convulsions of nearly the entire body. According to Ford[71] the patient develops delirium, has hallucinations and dies in coma. Ford appears to have based his discussion of symptoms in part on the well-publicized poisoning death of Count de Vecchi (see below), and the writer has found no other account in which such violent symptoms have been described.

Mycetismus nervosus is caused by ingesting fungi which contain muscarine (and related compounds) which are reputed to act upon nerve terminals. Such fungi are *Amanita muscaria* (Figure 63) which will also be discussed in the section on hallucinogenic fungi, a close relative, *A. pantherina, Clitocybe illudens* (Figure 64) and *Inocybe* spp. (Figure 65). Ford[72] listed *Inocybe infelix, I. infida, I. lateraria, I. sambucina, I. frumantacea,* and *I. repanda* as causing mycetismus nervosus as well as another species of *Clitocybe, C.*

FIGURE 63. Basidiocarps of the Fly Agaric, *Amanita muscaria.* Note annulus on stipe of basidiocarp at left, bulbous base (two center basidiocarps), and remains of universal veil on pilei. (Photograph by A. H. Smith.)

FIGURE 64. The Jack-o-Lantern, *Clitocybe illudens.* (Photograph by G. F. Atkinson.)

sudorifica. Apparently the species most frequently involved is the Fly Agaric, *Amanita muscaria,* but there exist today very considerable differences of opinion as to whether or not this is truly a poisonous species. One opinion is that *A. muscaria* is not lethal unless the patient has organic heart trouble, another is that it is merely hallucinogenic, and Wasson[192-195] makes a very strong case for his thesis that this well-known mushroom species in reality is Soma, the only plant ever deified by man. Even if it is proved that early Aryans did deify this mushroom, there is no guarantee that it is not poisonous, because there were some very capricious individuals among man's early gods. The facts are that at this point in time, in spite of the great upsurge in interest in this species during the last decade or two, there is presently not enough information based upon experimentally derived data to permit making any specific and exact statement regarding the toxicity and/or hallucinogenic capabilities of this and related species. There is even some question as to whether or not the fungus called *A. muscaria* in the U. S., which is usually seen with a yellow to orange pileus (covered with white warts), is the same as the Fly Agaric of Europe and Asia, which commonly has a bright red pileus (also covered with white warts).

A variety of related compounds such as muscaridine, muscimol, ibotenic acid, and muscazon (Figure 66) have been isolated from *Amanita muscaria* in addition to the muscarine that was isolated earlier and whose chemistry is now well known (Eugster,[63,64] Eugster and Waser,[66] and Wilkinson[201]). Eugster[65] has presented muscarine analyses for 31 species of fungi, some of which contain more than 1% (dry weight basis) of this compound. Catalfomo and Eugster[33] reported that muscarine isomers are predominant in some species of *Inocybe,* and

FIGURE 65. The Unfortunate Inocybe, *Inocybe infelix.* (Photograph by G. F. Atkinson.)

Genest et al.[77] reported that muscarine occurs in *Clitocybe illudens* (usually considered as poisonous) as well as three other species of this genus: *C. dealbata, C. cerussata,* and *C. rivulosa.* Reference to Table 2 shows that five of the eight authors whose reports are included in this table list *C. illudens* as poisonous. However, there is only one report each for *C. dealbata* and *C. rivulosa* and none of the eight authors list *C. cerussata* as a poisonous species. Even more interesting is the fact that without the benefit of the great number of chemical analyses that have been performed in relatively recent years, Ford was able in 1923 to group *Clitocybe* spp. and *Inocybe* spp. with *Amanita muscaria* solely on the basis of similarity of symptoms exhibited by individuals who had ingested them.

Exactly how muscarine is synthesized in *Amanita muscaria* has not been established, but it was predicted (e.g., Leete[113]) that a hexose is the direct precursor. However, Brown et al. (cf. Catalfomo and Eugster[33]) concluded that in *Inocybe napipes* the precursor was not D-glucose. There are a number of stereoisomers of muscarine and apparently they differ in their pharmacologic potency. Muscimol, muscazon, and ibotenic acid are related compounds, since muscimol can be formed from ibotenic acid by decarboxylation and loss of water, and muscazon can be formed from ibotenic acid by ultraviolet radiation (Eugster[64]). On the basis of the low concentration of muscarine and the relatively weak activity of this compound when taken orally, Eugster[64] concluded that muscarine can only be considered as a minor active component of *A. muscaria.* Fraser[76] administered muscarine chloride to monkeys orally and found that there was little effect even when the amount given was many times that which causes poisoning by the ingestion of *A. muscaria* by the human being.

Wieland et al.[199] reported isolating considerable quantities of the hallucinogenic indole compound, bufotenine, from *Amanita mappa*, and detecting it in small amounts in *A. muscaria* and *A. pantherina*, but Eugster[64] stated that it was not present in Swiss and South German varieties of *A. muscaria*. Puharich[145] stated without offering any evidence that bufotenine is present in *A. muscaria*, but Catalfomo and Eugster[34] stated unequivocally that this compound does not exist in this mushroom. Anyone reading the carefully documented accounts of Eugster and his associates and the rather bizarre and fanciful book of Puharich would probably be inclined to accept the statement of Catalfomo and Eugster. These latter workers also stated that bufotenine is inactive when administered orally.

Muscarine stimulates the parasympathetic nervous system and Simons[165] stated that fatalities from it would be more frequent were it not for the fact that it is poorly absorbed from the intestinal tract. Simons also stated that few deaths have been recorded from *Amanita muscaria* or *A. pantherina* poisoning (Buck[29] listed one death from *A. pantherina* poisoning during the years 1924 to 1961) and these mainly involved children or persons already organically ill. Webster and Pickering[196] reported treating a child with convulsions so severe that they could be controlled only by the administration of what appear to be

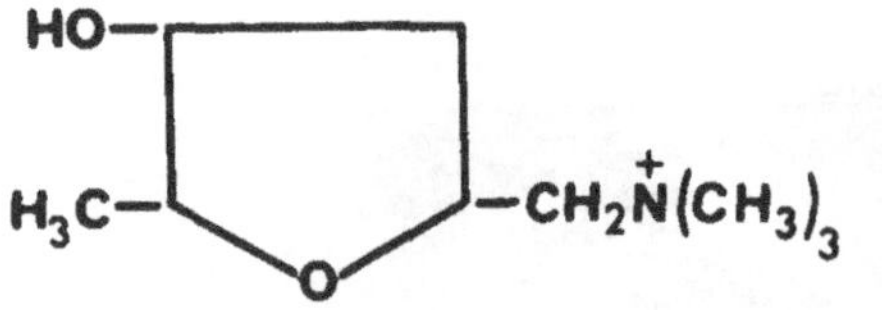

Muscarine

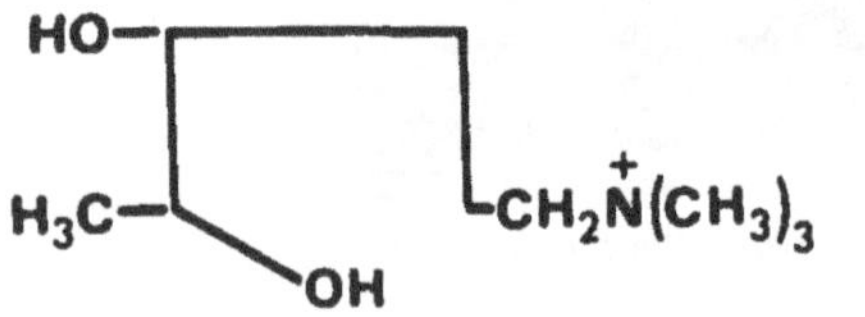

Muscaridine

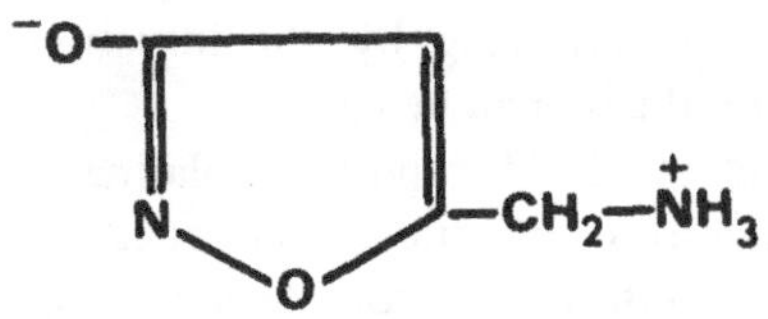

Muscimol

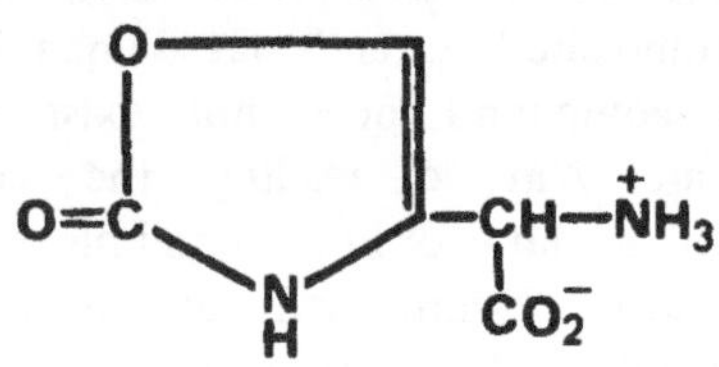

Muscazon

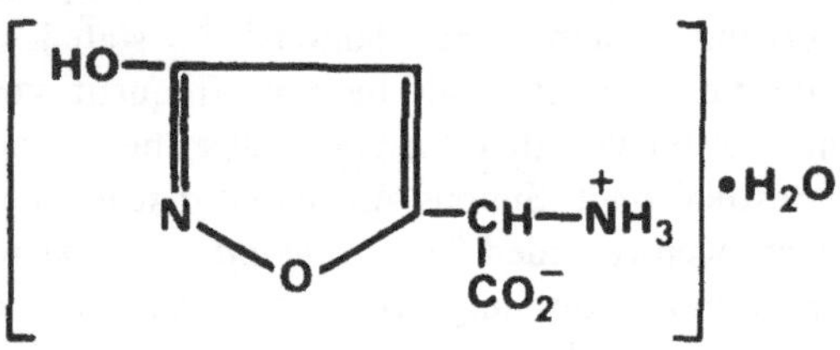

Ibotenic Acid

FIGURE 66. Some compounds that have been isolated from *Amanita muscaria, A. pantherina, Clitocybe illudens,* and *Inocybe* spp.

very large doses of curare and by intermittent positive pressure respiration.

In spite of the above discussion which in part seems to imply that perhaps *Amanita muscaria* is not toxic or at least has only limited toxicity (see also section on Hallucinogenic Fungi and Ethnomycology), there occurred in the U. S. one fatality which can almost surely be ascribed to the ingestion of this mushroom. The case was highly publicized and may in large measure be responsible for the rather widespread belief in the U. S. that the Fly Agaric is a highly lethal species. The particular case in question involves the fatal poisoning of Count de Vecchi, an attaché of the Italian legation in the U. S., in November of 1893 (Coville,[46] Prentiss,[144] Fischer[68]). Count de Vecchi purchased a considerable number of *Amanita muscaria* basidiocarps from a Virginia farmer on the assumption that he was purchasing the edible and tasty *Amanita caesarea* (Caesar's Mushroom; Royal Agaric; Orange Amanita), with which he apparently was not as familiar as he thought. He ate two dozen (!!!) mushrooms for breakfast which he finished by 8:30 a.m. By 9:00 a.m. he was in a state of collapse and soon lost consciousness. He had such violent convulsions that the bed was broken and he died the evening of the following day without regaining consciousness. A doctor who had breakfasted on only one dozen (!!) of the mushrooms also became unconscious by about 9:00 a.m. and remained so for five hours. The doctor suffered no pain but on the contrary was in a comfortable sleepy state (Soma?) and by 7:00 p.m. was out of danger. With all due respect to the Count it must be suggested that in fairly large measure his death might well be ascribed to simple gluttony. The fact that the doctor ate one dozen basidiocarps of *Amanita muscaria* and survived is certainly substantial evidence that if this species is generally toxic, its toxicity in no way approaches that of *Amanita phalloides*, one basidiocarp of which is often enough to kill. Again, with all due respect, it must be stated that if the dozen basidiocarps which the good doctor ate were at all close in size to the specimen of *A. muscaria* shown in Figure 2, gluttony may also have been involved in this unpleasant incident. On the other hand, the possibility that one or more basidiocarps of some species other than *A. muscaria* (e.g., *Amanita brunnescens*, Figure 61) was included with the mushrooms eaten, must not be precluded.

The entire question of *Amanita muscaria* (and other species containing muscarine and related compounds) poisoning is in a somewhat confused state at present and is obviously in need of further study. There is some possibility that the mushroom which is called *A. muscaria* in the U. S. is different from the European form; this point will be discussed at greater length in the section on Hallucinogenic Fungi and Ethnomycology. At the present time it appears that these fungi contain both toxic and hallucinogenic compounds, and with the growing interest in hallucinogens we can confidently expect considerable increases to be made in our knowledge of this subject in the near future. The principal toxic compound may be muscarine and the psychoactive compounds muscimol, ibotenic acid, and muscazon, although there may be still other active compounds which thus far have eluded isolation. According to Simons,[165] muscazon is least active and has not been isolated from *A. pantherina*, a close relative of *A. muscaria*. Theobald et al.[176] and Waser[187] have investigated the effects of muscimol and ibotenic acid and neither compound was found to produce the typical hallucinogenic syndrome of LSD, psilocybin, or mescaline. Scotti de Carolis[160] found that rabbits treated with muscimol showed EEG patterns different from those produced by other hallucinogens.

Atropine has long been considered to be a specific antidote for muscarine poisoning and when muscarine was thought to be the only physiologically active compound in *Amanita muscaria*, the administration of atropine was routinely recommended. Thus, Dreisbach[55] recommends as treatment after ingestion of *A. muscaria*: (1) removal of ingested mushrooms by gastric lavage or emesis followed by catharsis, and (2) the administration of 2 mg of atropine subcutaneously every 30 minutes until symptoms are relieved. However, atropine increases the toxic effects of muscimol and ibotenic acid (Simons[165]) and hence should not be administered if the fungus ingested contains these compounds. Thus, atropine is recommended in cases of poisoning by species of *Inocybe* and *Clitocybe* but not in connection with poisoning by *A. muscaria* or *A. pantherina*. This strikingly emphasizes the extreme importance of the attending physician in any case of mushroom poisoning knowing exactly what species of mushroom was ingested so that he may know what type of treatment is indicated. Some claims have been made that certain tropane alkaloids occur in *A. muscaria* but the evidence to date is not conclusive and Eugster[64] has criticized these claims.

4. **Mycetismus sanguinareus** – According to Ford[71] the only fungus capable of causing this type of poisoning is *Helvella* (*Gyromitra*) *esculenta*. However, *Helvella infula* is also listed as a poisonous species in Table 2, and Simons[165] and Mullendore[131] both list *Helvella underwoodii* as poisonous. As early as 1928, Seaver[162] denied that the name *Gyromitra esculenta* was valid, and many workers today place this name in the synonymy of *H. esculenta.* These "species" badly need critical study, since the typical ascocarps of *Gyromitra esculenta* that I have seen in ravines near the terminal moraine of the Wisconsin Ice Sheet in west-central Indiana usually do not resemble the illustrations and descriptions of other authors for this species. Seaver's illustrations of *H. esculenta* certainly do not resemble the specimens I have collected many times in Indiana and Ohio. For example, Simons[165] stated that the pileus is thin, wrinkled, folded, convoluted, and *usually somewhat saddle-shaped* - a description that could scarcely be applied to the ascocarp shown in Figure 67, especially with regard to pileus shape. Perhaps *Helvella esculenta* is an extremely variable species and what has been called *Gyromitra esculenta* is but one variant, but it would seem that the possibility exists that the huge ascocarps found both in Indiana and Ohio (up to 8 in. tall and with white, rugose stipes up to 4 in. in diameter) may not be *Helvella esculenta* at all. For this reason, this fungus will be referred to here as *Gyromitra esculenta*.

A curious feature of man's attitude toward the False Morel, *Gyromitra esculenta*, is that it is deliberately sought in the spring (it occurs in the same season and often in the same habitats as the true Morels) by many individuals and is considered by them a delicious mushroom, comparable to the true Morels. By others it is considered poisonous and there are known instances of individuals being fatally poisoned by this species (Buck[29] listed seven American fatalities as being due to the ingestion of this species during the years 1924 to 1961). According to Simons[165] the species is widely eaten in Michigan and Idaho, and the present writer has known many individuals in southern Illinois who seek out this mushroom in season. Some canned *G. esculenta* is imported

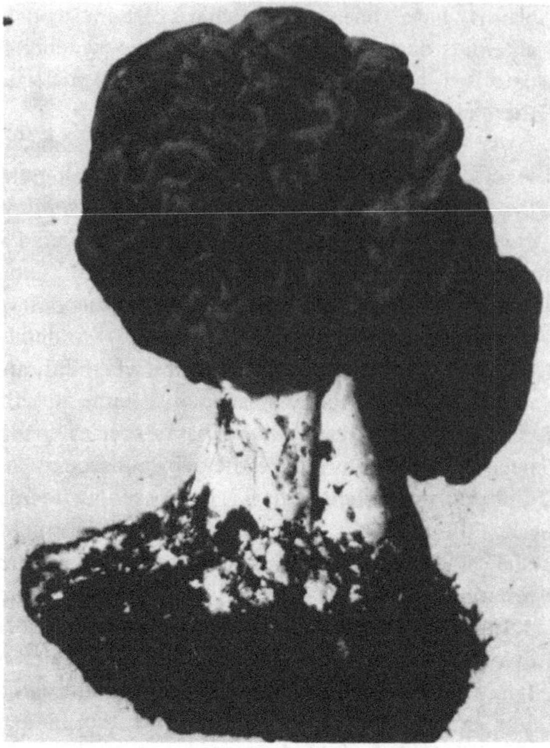

FIGURE 67. A poisonous False Morel, *Gyromitra (Helvella) esculenta.* (Photograph by D. P. Rogers.)

from both France and Sweden and so may occasionally be found in this form in the U. S., and Wasson[190] stated that when dried, this species is not poisonous and that it is sold in immense quantities in the markets of Europe. Simons attributed the low poisoning frequency of the species to the fact that in preparing it for eating most people boil it for a long time in large amounts of water and then discard the cooking water. Since, if *G. esculenta* can be collected, the true Morels can usually also be found, it would seem advisable not to attempt to eat this species but to rely on the *Morchella* species which are in season at the same time.

In mycetismus sanguinareus, nausea and vomiting which may or may not be accompanied by diarrhea commonly occur within 6 to 8 hours after ingestion (Ford[71]). Since the toxic principle of *Gyromitra esculenta* is hemolytic in its action, a determination of methemoglobin levels may provide an index to the degree of poisoning (Buck[29]).

Poisoning by *Gyromitra esculenta* has on occasion presented some rather confusing aspects. For example, only one of several persons who eat this mushroom at the same time may become seriously ill (Hendricks[92]), or a first meal may cause no ill effects while a second may result in poisoning (Dearness,[52] Mittman[129]). In the cases described by Hendricks, 8 people, ranging in age from 2 to 80 years, all ate False Morels. One, a woman aged 69, died in less than 5 days. Her daughter, aged 38, suffered from vomiting but recovered the next day; the remaining 6 suffered no ill effects. Dearness described a situation in which a family ate two quarts of *G. esculenta* for supper without any ill effects. The following day they ate one and one half quarts for lunch and that night all were ill. One died two days later, but the others recovered without any medical treatment.

In 1885, Boehm and Külz[25] isolated an acid from *Gyromitra esculenta* which had the empirical formula $C_{12}H_{20}O_7$ and which they named "helvellic acid." Ford[71,72] and many others believed that the poisonous principle of the False Morel was this heat-resistant helvellic acid. However, helvellic acid is now known to be a harmless mixture of fatty acids, and List and Luft[117,118] have recently identified the toxic compounds of *G. esculenta* as monomethylhydrazine and its derivatives, which do not exist in the free state in fresh ascocarps. These toxic compounds are released by hydrolysis, probably of a glycoside (Simons[165]). List and Luft named the first hydrolysis product gyromitrin (Figure 68). Thus, the amount of gyromitrin and subsequently the amount of monomethylhydrazine formed could well be a function of the time devoted to the parboiling operation.

Since monomethylhydrazine is used as a rocket fuel, its toxicity has been extensively studied by the U. S. Air Force (Clark et al.,[40] Patrick and Back,[140] Pinkerton et al.,[143] and Reynolds and Thomas[150]). It is a very toxic compound, the exact mechanism of action of which is not understood. Primarily it affects the central nervous system, but there is some action on other organs. Apparently the inactivation of B_6 vitamers is involved in its action. Back and Pinkerton[11] obtained results which may throw some light on the confused picture resulting from the observations of Hendricks[92] and Dearness[52] because they found that with Macaque monkeys the difference between a no-effect dose and a lethal dose of monomethylhydrazine is very small.

As treatment for poisoning by a mixture of hydrazine and dimethylhydrazine, Azar et al.[10]

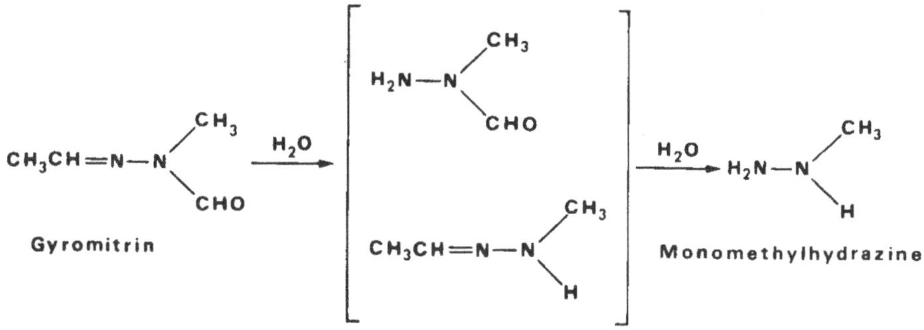

FIGURE 68. The toxic principle of *Gyromitra esculenta* is monomethylhydrazine, a hydrolytic product of gyromitrin, which presumably is a hydrolytic product of a glycoside.

recommended the administration of pyridoxine hydrochloride (25 mg/kg) and sodium phenobarbital. However, Mittman,[129] in treating a child who had been poisoned by *G. esculenta*, used thioctic acid at a level of 150 mg per day and reported a surprising response and recovery.

In spite of the fact that this mushroom is eaten by a great many people and marketed in both the canned and dried form in Europe (and to a small extent in this country), it is my opinion that this and related species should be avoided. Franke et al.[75] compiled reports of poisoning by this species (mostly in Europe) from 1782 to 1965. Their data on the total number of individuals who ate this mushroom are incomplete, but of 552 persons who became ill, 104 subsequently died, a fatality rate of almost 19%.

5. **Mycetismus cerebralis** – According to Ford,[71] four or five hours after the ingestion of mushrooms that cause mycetismus cerebralis, patients begin to show peculiar symptoms. In many ways they appear to be intoxicated: they are greatly exhilarated, laugh on the slightest provocation, have disturbances of vision, and develop a staggering gait. Ford listed *Panaeolus campanulatus* (Figure 69) and *P. paplionaceus* (Figure 70) as the mushrooms primarily responsible for this type of poisoning. The symptoms eventually disappear and the patients are usually restored to health in 24 to 48 hours. The report of Verrill[185] definitely implicated *P. paplionaceus* in a case of mycetismus cerebralis, but Wasson[190] denied that *P. campanulatus* can be similarly implicated and stated that "many of us have tried this species and found it innocent"; however, Child[36] reported that a subject became intoxicated after eating 2 g of this species. Kauffman[106] listed *Panaeolus retirugis* (Figure 71) as suspected of being poisonous and Herbert (cf. Trotter[178]) noted that a fourth species of *Panaeolus, P. ovatus,* had caused mushroom intoxication in New South Wales and was there called the "hysteria fungus."

FIGURE 69. *Panaeolus campanulatus,* which was listed by Ford[71] as one of the mushrooms causing mycetismus cerebralis, so it may be hallucinogenic. (Photograph by G. F. Atkinson.)

At the time that Ford considered and named this fifth category of mushroom poisoning, little was known in this country of hallucinogenic mushrooms and so he regarded the unusual symptoms described above as representative of a specific type of poisoning. In the light of more

FIGURE 70. *Panaeolus paplionaceus,* which, like its relative *P. campanulatus,* may be hallucinogenic. (Photograph by G. F. Atkinson.)

FIGURE 71. *Panaeolus retirugis,* which is listed by Kauffman[106] as a poisonous species. (Photograph by G. F. Atkinson.)

modern studies it now seems apparent that he was probably dealing with hallucinogenesis rather than toxicity. Whether the term mycetismus cerebralis should be retained as descriptive of a type of mushroom poisoning will depend upon whether or not the hallucinogenic compounds found in fungi of this type are shown to be toxic or if toxic compounds are found to be present in addition to the hallucinogenic ones. This type of "poisoning" will be discussed in the section on Hallucinogenic Fungi and Ethnomycology which follows.

Alcohol-mushroom Poisoning

In addition to the five categories of mushroom poisoning recognized by Ford and described briefly above, still another type of what may be poisoning should be mentioned here, although at this time there is no general agreement as to whether this type of poisoning exists. There have been a number of reports (Josserand,[104] Pilát,[142] Ramsbotton,[147] Smith,[171]) that a mild type of poisoning may occur when the edible Inky Cap, *Coprinus atramentarius* (and possibly other species of *Coprinus*), is ingested along with alcohol or if alcohol is taken shortly before or after eating this mushroom. In some ways this type of poisoning appears to resemble the "Antabuse®" (alcohol-disulfiram) reaction in which the individual is sensitized to alcohol. Simandl and Franc[164] reported the isolation of disulfiram (tetraethylthiuram disulfide) from basidiocarps of *Coprinus atramentarius* which seemed to point to an identical type of reaction as that involving Antabuse®. However, Wier and Tyler[200] were unable to find disulfiram in laboratory cultures of this species. There is obvious need for further experimental work in this area, although the disparate reports of Simandl and Franc and Wier and Tyler need not necessarily be viewed as conflicting, since it is quite possible that compounds might be found in the reproductive structure of a fungus which conceivably might not be found in the vegetative mycelium. Child[36] denied that *Coprinus atramentarius, C. micaceus,* or *C. comatus* can sensitize man to alcohol. He tested each species at least five times on the same subject, using raw, fried, or parboiled mushrooms in amounts of 100 to 600 g. Alcohol was taken during or after each meal and there were no ill effects. However, when the subject ate 2 g of *Panaeolus campanulatus* in a separate test he became intoxicated without alcohol. Child then suggested that what has been called alcohol – *Coprinus* poisoning might have been due to the accidental inclusion of *Panaeolus.* It is unfortunate that Child's tests were conducted with only one subject, since the subject may have been insensitive to the mushrooms.

Groves[84] reported one case in which two adults became ill when they drank rye highballs four hours after they had eaten fried morels but showed complete recovery after about four hours. Two children who had eaten the mushrooms at the same time but subsequently drank soft drinks showed no ill effects (there must be a moral here somewhere). Similarly, Shaffer[163] reported that three persons became ill when they drank highballs as they ate *Pholiota squarrosa* (Figure 33) which had been fried in butter. The causes of these types of poisoning have not been established, but Simons[165] has pointed out that they might result from increased solubility of some normally unassimilable substance.

Summation

Although a considerable number of fleshy fungi (largely agarics) have been variously reported by one or more authors to be poisonous, the fact still remains that nonpoisonous fleshy fungus species far outnumber the poisonous ones. In all probability the number of poisonous species (76) listed in Table 2 should be reduced because many have been cited by only one author who could have been in error regarding the identity of the species, and also because some instances of so-called poisoning may have been due to the ingestion of spoiled mushrooms or may have been an allergic reaction. The possibility also exists that some additional species may have to be added to the list, since the present day intensive search for hallucinogenic compounds among the fungi may well lead to the discovery of other poisonous species. This likelihood will become especially great if members of the modern "drug cult" start experimenting with unknown (to them) mushrooms in the hope of discovering one that is hallucinogenic.

Extant reports do not presently support Ford's early categorization of mushroom poisoning into five types. Thus, until specific toxic compounds are demonstrated to be present in the mushrooms said to be responsible for what Ford called mycetismus gastrointestinalis, this should probably no longer be recognized as a type of poisoning. On the contrary, since some individuals are "poisoned" and some are not, it would seem more logical to simply view mycetismus gastrointestinalis as a food allergy and recognize that there is a wide range in sensitivity of different individuals to different foods.

There is no question regarding the validity of considering mycetismus choleriformis (caused by ingestion of *Amanita phalloides* and related species of *Amanita,* as well as certain *Galerina* species), mycetismus nervosus (caused by *Amanita muscaria, A. pantherina, Clitocybe illudens, Inocybe* spp.), and mycetismus sanguinareus (*Gyromitra esculenta,* etc.) as specific and different types of poisoning, since definite toxic compounds have been isolated, their chemical structures established, and their effects studied. There is some confusion still surrounding poisoning of the mycetismus nervosus type, since in some of the species involved in this type of poisoning both toxic and hallucinogenic compounds have been isolated.

Ford's fifth category (mycetismus cerebralis) probably should not be further recognized as mushroom poisoning or at least only tentatively so until it can be determined whether fungus hallucinogenic compounds are either toxic or are accompanied by toxic compounds. The symptoms listed by Ford are remarkably similar in some respects to those described by some individuals who have deliberately ingested "the sacred mushroom" of Mexico and Central America. Oddly enough, in spite of the great outcry in some areas against the use of hallucinogens such as peyote and LSD, nothing has been said about outlawing the hallucinogenic mushrooms. This is undoubtedly merely another reflection of the general lack of knowledge concerning the fungi rather than any lack of vigilance on the part of the anti-hallucinogen interests.

Poisoning by certain species of mushrooms (*Coprinus atramentarius* has usually been blamed) only when they are ingested along with alcohol needs further investigation. There has been only one report of the isolation of disulfiram (Antabuse®) from mushrooms and until this report is verified by other workers or other specific toxic or sensitizing compounds have been isolated, the mere fact that someone occasionally becomes ill does not necessarily mean that poisoning is involved. It should also be borne in mind that a not inconsiderable number of individuals become ill when they drink and eat, regardless of what type of food they ingest. In any event, poisoning in such instances is not due solely to the mushrooms – alcohol combination and, at least in some instances, may be due to the alcohol alone.

While animal feeding experiments may have some value as a means of determining the toxicity of mushrooms to humans, this entire area needs rather critical inspection. The fact that an experimental animal may be susceptible or immune to the action of a specific mushroom toxin does not necessarily guarantee that man will react to the same toxin in the same way. Thus, Fraser[76] was able to administer relatively large quantities of muscarine to monkeys with little adverse effect, and Miller and Halls[128] (see section on Fungi as Animal Food) listed six mushroom species, reported to be poisonous to man, as food utilized by a variety of species of wildlife. Furthermore, not all animals react the same way after ingesting poisonous mushrooms. Limousin and Petit[116] found that *Amanita phalloides* was toxic to cats but not to rabbits, a finding which may in part be responsible for the often expressed opinion that carnivorous animals are generally more susceptible than herbivorous animals to mushroom poisoning.

From the standpoint of number of fatalities the really serious type of mushroom poisoning is that caused by *Amanita phalloides* and its various close relatives such as *A. verna, A. virosa,* and *A. bisporigera.* The fact that *A. phalloides* does not appear to be very common in the U.S. should not give anyone a false sense of security, because its white relatives are very common and from all accounts are equally toxic. For many years there were no reliable antidotes against the toxins of these highly poisonous mushrooms; however, recent findings indicate that thioctic acid may have real promise in the treatment of individuals who have become poisoned. In spite of the fact that the Royal Agaric (*Amanita caesarea*) is an edible mushroom and is considered a delicacy by many knowledgeable individuals, no species of *Amanita* is recommended for the beginning mycophagist. In fact the beginner should learn to recognize *Amanita* and avoid it for eating purposes, regardless of species. Perhaps it might be well to bear in mind that Count de Vecchi died from eating what he thought was the Royal Agaric.

HALLUCINOGENIC FUNGI AND ETHNOMYCOLOGY

It would be unrealistic today to speak in broad terms about the eating of wild mushrooms without giving some space to the related subjects of hallucinogenic fungi and the newly developing field of ethnomycology. Three decades ago it would have been possible to write a short paragraph or two on the subject of the use of hallucinogenic mushrooms in Siberia, but one would have been hard put to find much evidence at that time that mushrooms had ever been very much involved in man's culture and in his religious practices. Today, thanks to the pioneering researches of the Wassons, who are actually the founders of ethnomycology, and the contributions of a growing number of other workers, we are beginning to see more sharp outlines of what may be a somewhat different picture. It is no longer possible to view mankind's early contacts with mushrooms solely in terms of food gathering or the avoiding of foods which might be poisonous. For the individual specifically interested in this fascinating and exciting aspect of mycology, the early book of the Wassons[189] and the more recent *Soma, Divine Mushroom of Immortality* (Wasson[193]) are indispensable, although extreme luck would have to be involved in finding the earlier work, which was printed in very limited edition.

Geographic Distribution of the Practice of Hallucinating with Fungi

The eating of hallucinogenic mushrooms for ritual, orgiastic, or other purposes is to be found in several different places in the world, often far removed from each other, and there is no substantial evidence that these far-flung practices are in any way related to each other.

1. **Siberia** – Unless it can definitely be established that the Soma referred to so frequently in the RigVeda was *Amanita muscaria,* the first reference to the use of hallucinogenic fungi was in the diary of Kamieński (cf. Wasson[193]), a Polish prisoner of the Czar in Siberia, who in 1658 wrote that he saw members of the Ob-Ugrian Ostyak tribe getting drunk on *Amanita muscaria.* There followed a variety of reports of other observers, which served to establish that the practice of eating this mushroom as an intoxicant or hallucinogen was one that was shared by several primitive Siberian tribes. On the basis of these not too objective early observations it appears that the details of the practice were not identical among all of the tribes, since Donner[54] reported that in the Ket and Selkup tribes it was believed that only shamans or individuals who were going to become shamans could eat the Fly Agaric with impunity and that others would die if they ate this mushroom. Although specific details of the practice among the various Siberian tribes have not been described, since the Soviet government has not been inclined to permit competent anthropologists from other countries to visit and study these primitive peoples, there is no question but that it was rather widespread at least among the peoples of the Kamchatka area. Two Soviet scientists (Brekhman and Sem[28]) have recently reported that the minor nationalities of Siberia and the Far East do not now use any psychoactive drugs but admitted that earlier such peoples as the Itelman, Koryak, Chukchi, and Yukagir ate *A. muscaria.* Apparently when the Russians occupied Siberia they introduced vodka as a substitute for the Fly Agaric with the double motive of suppressing the use of the latter and making a profit at the same time.

One curious feature of *Amanita muscaria* which was commented on by many of the earlier travelers is that when it is eaten the hallucinogenic principles are said to be excreted through the urine, and for the poor individual who had no mushrooms, drinking the urine of someone who had eaten such mushrooms achieved the same general results. To most travelers who observed such urine drinking, the habit was extremely disgusting and may have led to their viewing the general practice of mushroom eating somewhat less objectively than might be desired by modern anthropologists and ethnomycologists. Wasson[193] noted that reindeer are extremely fond of both *A. muscaria* and human urine but to date it has not been established that this fondness is necessarily because of the hallucinogenic properties of these items.

2. **Europe** – In general there seems to be little evidence that hallucinogenic mushroom cults ever developed in Europe. Thus Wasson[193] stated that such mushrooms were not involved in the witch-

craft epidemics which plagued this continent during Medieval and Renaissance times, and that he had found no reference to mushroom usage in the records of the Druids. There is a recurrent story that the Vikings known as Berserks went "berserk" as a result of eating *Amanita muscaria,* and I am among others guilty of repeating this story (Gray[79]). However, Wasson is of the opinion that the Berserks did not use this mushroom because it is not so noted in the Sagas and Eddas and because the symptoms of berserk-raging are not those of Fly Agaric inebriation. Perhaps it is time to remove this story from the literature or at least regard it as only tentative until further evidence is at hand.

Another item which has raised some question as to the possibility of a mushroom cult having been linked with early Christianity in Europe is the existence of an unusual fresco in the Plaincourault Abbey in France. In this abbey which was built in the late 13th century is a fresco (Figure 72) which shows Adam and Eve in the Garden of Eden. The curious feature of this fresco is that the Tree of Knowledge which occupies the central position resembles a huge *Amanita muscaria* basidiocarp with an unusual tripartite upper stipe. The "stipe" is bulbous at the base and there are four smaller "mushrooms" clustered around the large central one. This "tree" has been interpreted numerous times as representing the Fly Agaric, thus providing a somewhat vague association between hallucinogenic mushrooms and early Christianity. However, according to Wasson,[193] art historians do not agree with the mycological interpretation and state that the "tree" is a conventionalized method of representing a tree which was prevalent in Romanesque and Early Gothic art.

Puharich[145] makes a curious attempt to imply indirectly that the megalithic stoneworks sometimes known as cromlechs may have been representations of mushrooms. He states that the purpose of these cromlechs (more properly termed portal dolmens) is not known, but this is certainly not the opinion of English and Irish archeologists. Portal dolmens are found in Ireland, England, the European continent, Africa, Anatolia (Alkim[3]), and the Malbar coast. Undisturbed dolmens which have been excavated in Ireland and elsewhere have usually revealed burials, and many archeologists believe that after the burial was completed a

FIGURE 72. The Fresco of Plaincourault, Plaincourault Abbey (1291 A.D.). (From *Soma, Divine Mushroom of Immortality.* By permission of R. Gordon Wasson.)

FIGURE 73. A portal dolmen (cromlech) from Carrowmore, County Sligo, Ireland. In this instance the capstone is hemispherical but this dolmen does not resemble a mushroom.

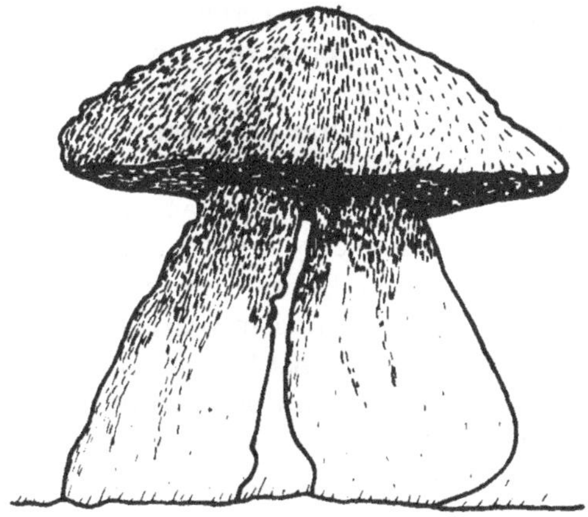

FIGURE 74. The Malabar Cromlech, Wiltshire. Artists during the Romantic period often made bizarre illustrations of prehistoric stoneworks such as portal dolmens. (Redrawn from an 1806 lithograph.)

dolmen was covered with a mound of earth which was often ringed with boulders; over the centuries the soil washed off, revealing the standing stones with their larger capstone (Figure 73).

I have seen many portal dolmens (there are large concentrations of them in County Sligo, Ireland) but have yet to see one which bears more than a passing resemblance to a mushroom, and the Anatolian cromlech figured by Alkim[3] in no way resembles a mushroom. Perhaps Puharich was deluded by having seen the sometimes rather fanciful illustrations of "cromlechs" which appeared during the Romantic period, such as that shown in Figure 74. Puharich may be correct in his belief that many dolmens were destroyed by Christians because of the pagan origins of these stoneworks (after all, the Sphinx is reputed to have been disfigured by overzealous Coptic Christians in Egypt), but there is no evidence that this paganism was in any way associated with mushrooms.

3. **Egypt** – At this point any suggestion that a significant mushroom cult developed in Pharonic Egypt must be relegated to the realm of pure fantasy. As far as can be determined the only proponent of such a view is Puharich,[145] whose only evidence is based on the "hieroglyphs" written by a subject who claimed to be extraordinarily sensitive in the area of extrasensory perception. Ra Hotep, a supposed descendant of the great Egyptian architect who is believed by serious and competent Egyptologists to have invented massive building in stone through his building of the step pyramid of Xoser at Saqqara (3rd Dynasty), is purported to have revealed the knowledge of an early Egyptian mushroom cult through the writings of Puharich's sensitive subject. While Imhotep was definitely an historical figure, he is a very shadowy one and little is known about him by Egyptologists; even less seems to be known about "Ra Hotep."

4. **Middle East** – Even more curious than Puharich's attempt to link a mushroom cult with Pharonic Egypt is the book by Allegro,[4] *The Sacred Mushroom and the Cross.* In this book the author apparently attempts to establish the thesis that Judeo-Christianity derived from a small cult of mushroom eaters. Allegro's position has few if any adherents and before reading his book the reader is seriously advised to read the thoughtful review of Jacobsen and Richardson.[101] Ingalls[99] was rather kind and referred to Allegro's book as a spoof and well it may be. However, if it was meant to be a spoof it is at best in poor taste, renders a disservice to the developing field of ethnomycology, and may be regarded as offensive by many people. Unfortunately, the book appeared shortly after Wasson's serious and well-documented book on Soma and the two may have been confused in the minds of some people.

5. **New Guinea** – At irregular intervals, what is known as "mushroom madness" strikes certain natives of the Wahgi Valley of New Guinea. Forty years ago the people of this region were still living in the stone age and have not been studied for long

periods of time as have certain other primitive peoples. Ross[154] reported that a wild mushroom called *nonda* makes the user temporarily insane and that it is used before going out to kill another native. Since that time several other reports on mushroom madness have appeared, and Singer[168] reported that a species of *Russula* was responsible. In 1963 Heim and Wasson[91] visited the Wahgi Valley for a short period in order to study this practice. They learned that Europeans living in the valley knew of outbreaks of mushroom madness and were frightened by them, that the natives believed that the mushrooms only affected certain individuals, that in men the effects led to abnormal physical activity while in women the effects were erotic, and that the natives claimed that seven mushrooms were responsible. These were identified as *Boletus reayi, B. kumaeus, B. manicus, B. nigerimus, B. nigroviolaceus, Russula delica,* and *Heimiella anguiformis.* Heim and Wasson found that no psychotropic mushrooms were in use in the Wahgi Valley but reported that the natives attributed their extraordinary behavior to mushrooms. As Schultes[159] has pointed out, the "mushroom madness" of New Guinea needs further study. Wasson[190] stated that hallucinogenic mushrooms have been used deliberately in Borneo but gives no further details.

6. Indus Valley – About 1500 B.C. a people who were called Aryans came from the north through Afghanistan and occupied the historic Indus River Valley. They were related to the people of the Iranian plateau and were probably responsible for the destruction of the early Indus Valley civilization centered at Mohenjo-Daro and Harappa (Woolley[203]), since their writings speak of the Aryan god Indra as the "destroyer of ancient castles," and the archeological findings at Mohenjo-Daro certainly show evidence of widespread destruction. The bulk of our information concerning these Aryans is derived from a collection of over 1,000 hymns (known as the RigVeda) composed by their priests and sung during their worship. Apparently there is no question among Vedic scholars but that the Aryans deified some sort of plant which was known as Soma; however, there have been and still are differences of opinion regarding the identity of this god-plant.

Evidence of the reality of Soma is provided by the fact that Sanscrit ritual texts describe in detail how it should be prepared. Wasson,[193-195] relying on the assistance of the Vedic scholar O'Flaherty,[137] has presented very persuasive arguments that the Soma of the RigVeda was in reality the Fly Agaric, *Amanita muscaria.* The Sanscrit scholar Ingalls[99] expressed the opinion that Wasson had correctly identified Soma but objected to Wasson's thesis that Vedic priests impersonated their gods (Ingalls[99,100]), a point that seems to revolve around the question of whether or not the priests urinated Soma. The Aryan Soma is associated with Iran through the fact that there are records of a substance known as *haoma* which was offered up to Yima, the first worshiper of Ahura Mazda.

The true identity of Soma has been lost in antiquity and unless a major breakthrough occurs in the research devoted to this area and new information is brought to light it may never be known with certainty. At this time Wasson's identification of *Amanita muscaria* with Soma seems a very plausible one. An objection to this thesis might be raised on the grounds that no surviving traces of a mushroom eating cult have been reported from India, but in view of the fact that rituals and religions change with time this does not seem a valid objection. Even if Wasson's thesis is proved to be wrong it has served a very useful purpose since for the first time it has been instrumental in bringing together the diverse disciplines of biochemistry, mycology, anthropology, linguistics, pharmacology, etc.

7. United States – There is no question but that several species of hallucinogenic mushrooms occur in the U. S. and it was pointed out in the previous section that Ford's[71] fifth category of mushroom poisoning (mycetismus cerebralis) should probably more properly be regarded as hallucinogenesis rather than mushroom poisoning. In the U. S., however, such incidents involving the ingestion of hallucinogenic mushrooms appear to have been the result of ignorance or accident, and there are no reports of the deliberate eating of known hallucinogenic fungi. Similar accidental ingestion of an hallucinogenic mushroom known as the hysteria fungus has been reported from New Zealand.

8. Mexico and Central America – The practice of using hallucinogenic mushrooms ritually is best documented, at least in recent times, from Mexico. The early Spanish conquerors and priests apparently were aware of the practice, and there are records of it in early Codices (Figure 75). Was-

FIGURE 75. A 16th century Mexican drawing which shows a clump of mushrooms, a man eating mushrooms, and a god in back of the man, speaking through the mushrooms. Originally published in the Magliabecchiano Codex (Codex #232, Biblioteca Nazionale, Florence, Italy). (From Wasson.[188] By permission.)

son[191] stated that at least 25 of the early sources refer to *teonanácatl* (God's flesh), the name applied to the sacred mushrooms of Central America, but there are no records of early experimentation with these fungi. The first white people to attend a mushroom rite did so in 1938 in Huautla, Mexico (Johnson[103]). Several years later three types of hallucinogenic mushrooms were identified as *Psilocybe caerulescens* var. *matzatecorum*, *Panaeolus campanulatus* var. *sphinctrinus*, and *Stropharia cubensis*. Heim later identified 14 species of *Psilocybe*, *Stropharia*, and *Conocybe* as well as several subspecies as being sacred mushrooms (Wasson[191]), and a considerable number of mushroom species are now considered to be hallucinogenic (Heim,[88,89] Schultes,[157,158] Singer,[167] and Singer and Smith[170]). Heim's colored illustrations of five species and two varieties were included in Wasson's[188] *Life* Magazine article. The Wassons learned in 1952 that a 16th century mushroom cult had existed in Mexico and spent several rainy seasons there searching for surviving traces of this cult. There Wasson participated in the mushroom eating ritual and the Wassons' first accounts of the Mexican sacred mushrooms appeared in 1957 (Wasson,[188] Wasson and Wasson[189]) but the full account has yet to be published. Lowy[121] described his own participation in such a ceremony and also compiled a general list of hallucinogenic mushrooms. He includes, of course, *Amanita muscaria* (but not as a Mexican sacred mushroom); oddly enough there is no report of *A. muscaria* being used in Mexico, although the red-capped type common to Europe is known to occur there. Lowy's list is as follows:

Amanita muscaria	*Psilocybe cordispora*
Boletus kumaens	*Psilocybe hoogshageni*
Boletus manicus	*Psilocybe mexicana*
Boletus nigerrimus	*Psilocybe mixacensis*
Boletus nigroviolaceus	*Psilocybe semperviva*
Panaeolus sphinctrinus	*Psilocybe wassonii*
Psilocybe acutissima	*Psilocybe yungensis*
Psilocybe aztecorum	*Psilocybe yungensis* var. *deconica*
Psilocybe bolivari	*Psilocybe zapotecorum*
Psilocybe caerulescens	*Russula delica*
P. caerulescens var. *mazatecorum*	*Russula nondorbingi*
P. caerulescens var. *nigripes*	*Stropharia cubensis*

Insofar as I am aware, there are no published reports of a present day mushroom cult in Central America, although it seems more than probable that such a cult may once have existed in Guatemala, or at least mushrooms may once have played a rather significant role in the culture of the early people of that country. While the Romans undoubtedly held certain mushrooms in very high esteem (principally because of their taste) as is evidenced by their writings, there is only one report of their having carved a mushroom likeness in stone (Figure 76). It would appear that the representation of mushrooms in stone reached its zenith in Central America some time between 200 B.C. and 300 A.D. although it was practiced before this time. An astonishing number of carved

FIGURE 76. An ancient mushroom carving in stone (Roman). From the ruins of Timgad, Algeria (founded 100 A.D., destroyed 534 A.D.). (From Harshberger.[86] By permission of *Mycologia.*)

mushrooms (thus far about 100) have been found, mostly in Guatemala although a few are known from El Salvador and Mexico. Apparently these "mushroom stones" are highly prized collector's items and since most of them seem to have been found and collected by individuals other than archeologists, their precise original locations are largely unknown. Thus, of the stones found to date only a few are known to be derived from burial sites, so one can only speculate regarding their use by the people who made them. De Borhegyi[26] reported on nine miniature mushroom stones that were found in a sealed cache in Kaminaljuyu, Guatemala. Since nine miniature metates (grinding stones) were found with these mushrooms, Wasson[191] suggested that they may have been used for crushing mushrooms. Lowy[121] is of the opinion that there remain many more mushroom stones to be found in Guatemala. Several such stones are shown in Figure 77.

Sapper[155] appears to have been the first to suggest that these rather common stone artifacts represented mushrooms, since before that time they were considered to be phallic objects, an all too common view taken of almost any unexplained elongated stone object derived from an ancient site. Kidder et al.[107] referred to them as objects of unknown use, although later Shook (see Lowy[121]) considered the possibility that they may have been used as land markers, a not unreasonable possibility in view of the carved and inscribed boundary stones which have been found in the course of the excavation of Babylonian and Sumerian sites. Such stones have also been thought to have been used as seats (Thompson[177]), but Lowy correctly points out that they would be rather uncomfortable ones by our standards.

The most plausible explanation of the use of these stones is that they had ritualistic significance in a mushroom cult that existed many years ago. Thus Wasson and Wasson[189] suggest that there may be some relation between such a cult and the surviving rite of the use of hallucinogenic mushrooms in Mexico. Similarly, Lowy[121] points out that their ritualistic significance was possibly associated with a mushroom cult that disappeared in Guatemala centuries ago, and that they may be relics from a Mayan past in which the ceremonial consumption of mushrooms was practiced. Figure 77E, showing a pregnant female, seems to also associate human fertility with mushrooms. Further circumstantial evidence that such stones had ritual use is provided by the fact that such artifacts have sometimes been found in pairs. Lowy[120] described his participation in a mushroom eating ceremony as recently as 1969. The hallucinogenic mushrooms which he ate were identified as *Psilocybe caerulescens* var. *mazatecorum,* and Lowy makes a point of the fact that the mushrooms were offered to participants *in pairs,* a practice which Wasson[188] had commented on when he described his participation in such a ceremony.

Thus, while this interpretation yet remains to be proved, the preponderance of evidence to date seems to indicate that these mushroom stones had ritualistic significance and that they may well have been associated with the ancient counterpart and possible forerunner of the ritualistic consumption of hallucinogenic mushrooms found today in Mexico. Perhaps the most curious coincidence of all is that the fresco (Figure 72) in a 13th century French abbey shows what resembles *Amanita muscaria* with a tripartite upper stipe, and mushroom stones with tripod bases have been found in Guatemala.

A not impossible explanation of the significance of these stones might be that they were purely decorative in character. The assignment of only ritualistic or utilitarian purposes to most of the artifacts recovered from the residues of ancient cultures seems to imply that an appreciation of artistic objects was poorly developed or even lacking. In the opinion of this writer such a view is extreme and does little justice to earlier peoples. If one examines the fabulous collection of articles that was removed from the tomb of the fabled

FIGURE 77. Guatemalan mushroom stones. Early preclassic – 2000 to 1000 B.C.; late preclassic – 300 B.C. to 300 A.D. A – animal effigy, late preclassic. B and C – anthropomorphic, early preclassic; D – human effigy, late preclassic; E – human pregnancy effigy, late preclassic. (Photographs courtesy of B. Lowy.[121] By permission of *Mycologia.*)

Tutankhamen (e.g., the shrine for canopic jars and the golden throne) there are many objects that had utilitarian or ritualistic use. On the other hand there are also many objects that were purely decorative in nature and probably were considered by the people of that time as art objects. Furthermore, many of the utilitarian and ritualistic objects are truly works of art, a fact which most certainly does not indicate a lack of appreciation for beautiful objects.

In this connection one further comment needs to be made here. During the past decade mushrooms have found increasing use as a motif in many types of art objects as well as a decorative motif in many utilitarian objects (Figure 78). Thus, very attractive printed textiles are covered with mushrooms (mostly unrecognizable species); oil paintings of clusters of mushrooms are offered for sale; ceramic, glass, wood, plastic, and wax mushrooms are produced by the thousands; and chessmen in the form of mushrooms are not uncommon. Upon the excavation of the more durable of these objects by archeologists 3,000 years hence, it would be unwise to postulate that a "mushroom cult" existed in the U. S. during the late 1960's and early 1970's. This is especially true

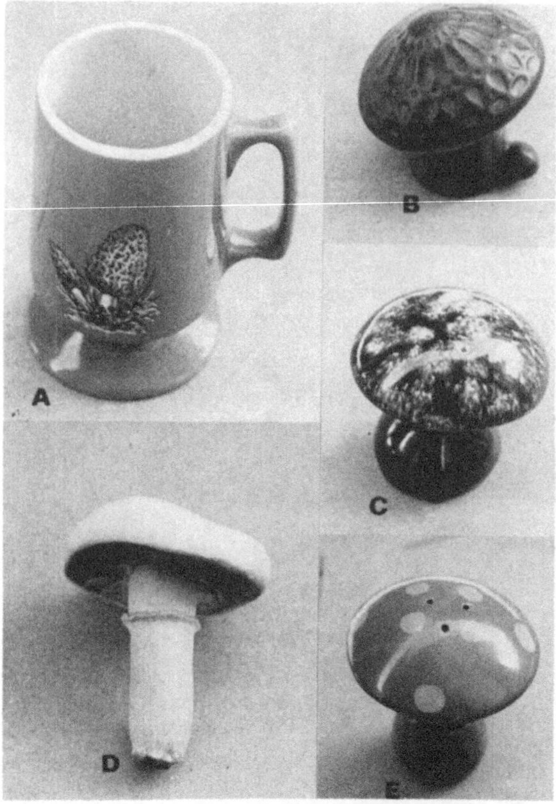

FIGURE 78. A modern mushroom cult in the U.S. Articles such as these are currently available in quantity in the U.S. They may be utilitarian (A, C, E), utilitarian and decorative (A, C, E), or purely decorative (B and D). A, C, and E are of glazed ceramic; B is of unglazed ceramic; D is of soft plastic. Also available are mushroom objects of wood, wax, metal, and fabric or various combinations. These are decorative, utilitarian or both, but what will the archeologist's interpretation of B be when he finds it in 5000 A.D.?

since there exist today many people who under no circumstances would knowingly eat a mushroom of any type, but who have no qualms about displaying a beautiful ceramic representation of *Amanita muscaria* as an art object on their desks or as a salt shaker on their tables. Curiously enough, representations of this species seem to be held in especially high esteem by the purchasers of such decorative objects.

Obviously there is a need for much more research in this area of archeology. It is to be hoped that Lowy is correct and that there are many mushroom stones yet to be discovered in Guatemala. But more especially it is to be hoped that competent archeologists will find these stones *in situ*. Perhaps then we will gain a much better understanding of the role that mushrooms played in the cultures of these early peoples. One cautionary note should be added. Since mushroom stones seem to be rather popular collectors items, if an increasing number of them should appear from sites unknown, some consideration should be given to the fact that a considerable traffic in modern made "ancient" Etruscan and Anatolian vases is known to exist!

Effects Induced by Hallucinogenic Mushrooms

1. *Amanita muscaria* – Since the early observations which were made on the effects of eating *Amanita muscaria* in Siberia were made by not too objective travelers who, with one exception, apparently made no attempt to eat these mushrooms themselves, no account of the symptoms they described will be presented here. Wasson[193] has discussed each of these early accounts and if interested, the reader should consult his book.

Wasson and three Japanese mycologists tried eating the Fly Agaric in Japan in 1965 and 1966 and with one exception reported disappointing results. They ate the mushrooms raw, drank the juice, and mixed the juice with milk and drank it, always on empty stomachs. They felt nauseated, and some vomited and felt disposed to sleep. They fell into deep sleep from which shouts could not rouse them. Wasson stated that once he had vivid dreams, but nothing happened like that which had happened when he ate *Psilocybe* mushrooms in Mexico. The one exception occurred when one of Wasson's Japanese associates ingested his mushrooms with a Japanese soup, having first toasted the pilei on an open fire. After awaking from the sleep that followed, he was elated and for three hours was a compulsive speaker. The substance of his remarks was that this was nothing like the alcoholic state but was infinitely better; no further details were given. Aside from the report of Puharich,[145] the writer has found no additional modern reports of the deliberate ingestion of *A. muscaria*. Puharich fed small amounts of this mushroom which he had collected in Massachusetts to 35 subjects, but none had hallucinations. It may be that the amounts fed were too small (in Siberia as many as ten were eaten), or it may be that the mushroom which is called the Fly Agaric in America is not the same species or strain as the European or Asian *A. muscaria*. These latter commonly have a brilliant red pileus with white warts, and while others have reported finding this red-capped type in the U. S., those seen by the present writer have always been yellow to orange. Heim[90] has raised the question of whether the American Fly Agaric is really *Amanita muscaria*.

With few well-documented modern accounts of the hallucinogenic effects of *Amanita muscaria* and the general belief in the U. S. that this is a toxic species, there is little wonder that not much experimentation has been conducted with the deliberate ingestion of this mushroom *per se*. However, the isolation and identification of such substances as ibotenic acid, muscimol, and muscazon (Figure 66) has made it possible to study the effects of these compounds separately with the hope of gaining a better understanding of the effects of *A. muscaria*. Thus, both Eugster[64] and Waser[187] stated that ibotenic acid is the principal active ingredient of *A. muscaria*, muscimol is next, and muscazon is least active. However, Catalfomo and Eugster[34] suggested that muscimol is probably an artifact generated in process, but stated that there is no doubt that it is psychomimetic. They stated that ibotenic acid and muscazon behave similarly but are somewhat less potent than muscimol. Since wide variations in the effects of *A. muscaria* have been reported, Eugster suggested that this might be the result of natural variations in the ibotenic acid – muscazon ratio.

Waser[187] reported on experiments done on himself with careful psychiatric control. When he ingested 20 mg of ibotenic acid in water, it tasted like mushrooms but produced little immediate action. Within half an hour his face was warm and slightly flushed but there were no changes in blood pressure and heart rate. There was no psychic stimulation but only lassitude followed by sleep. One day later lateral migraine with one-sided visual disturbances occurred; this continued in a milder form for two weeks. When Waser ingested 5 mg of muscimol it had little effect except to induce a feeling of laziness. Ingestion of 10 mg produced slight intoxication after 90 minutes with dizziness, ataxia and elevated mood, psychic stimulation and no hallucinations, but there were slight changes in taste and color vision. There was some myoclonic muscle twitching, then sleep with dreams. After two or three hours Waser felt normal; there were no further effects as there had been with ibotenic acid.

Catalfomo and Eugster[34] stated that present knowledge of the chemistry of *Amanita muscaria* cannot fully explain the physiological actions induced by ingestion of this fungus and suggested that it would be unlikely that any one compound would be found to be the sole causative agent. Although great steps have been taken in attempts to gain a thorough understanding of the chemistry of this mushroom, we are still a long way from understanding it, and the more adventurous reader is specifically advised not to experiment with it. At the present time it appears that, with respect to this mushroom, the whole may be considerably more than the sum of all of its parts.

Throughout the present work *Amanita muscaria* has been referred to repeatedly under the long-established common name of Fly Agaric, since it has been a common belief for many years that this mushroom could be used as a fly poison. Wasson[193] expressed doubt that it can kill flies. He noted that during the Middle Ages and the Renaissance, demons were represented in the shape of flies and that the word "fly" has been used in the past in a variety of other senses. Therefore, it is quite possible that the Fly Agaric acquired its name from something that had nothing to do with fly killing. Wasson stated that *A. muscaria* does not kill flies but merely drugs them, and Baxter (unpublished) was unable to kill flies with this mushroom. In an experiment without any controls, Puharich[145] claimed to have demonstrated the fly-killing ability of *A. muscaria*. This latter worker placed 10 flies under a bell jar with a large basidiocarp of *A. muscaria* and reported that 8 of them died within 24 hours. He did not state whether the flies ate the mushroom and from the manner in which his experiment was conducted a more plausible explanation of their death would be that they died from oxygen depletion resulting from the respiration of the large, fresh basidiocarp (stipe 8 in. long; pileus 6 in. in diameter). Ibotenic acid, one of the compounds isolated from *A. muscaria*, is known to have mild insecticidal properties but apparently has only limited use for this purpose.

As I have done, Wasson[193] has also expressed his abhorrence of the use of the word "toadstool" as applied to mushrooms and stated that this term has encapsulated the loathing and fear of English-speaking people for wild mushrooms. The term probably had its origin in the fact that for many years the toad was considered evil, and Wasson stated that the abuse heaped on the toad was probably one of the fruits of Christianity.

2. The Sacred Mushrooms of Mexico – The

effects of eating the "sacred mushroom" of Mexico are much better documented in modern times than those caused by eating the Fly Agaric. The world was generally unprepared for the reports (Wasson,[188] Wasson and Wasson[189]) in 1957 that an hallucinogenic mushroom-eating cult existed in the modern world. Even more startling was the fact that the mushroom-eating ritual as described by Wasson was mingled with Christian practices. Actually, this latter finding should not have been so surprising since there exist in the world today many examples of early Christianity having been grafted on paganism. As far as can be determined, Wasson and his associate Richardson were the first white men to eat the divine mushroom.

On June 29, 1955 Wasson and Richardson, along with about 18 Indians participated in the mushroom-eating ritual at Huautla in Mexico. The ceremony began about 8:00 P.M., the atmosphere was friendly but solemn, and it was conducted by a woman, the *curandera* or shaman. In some ways it resembled a communion, with chocolate first being served ceremonially. At about 10:30 P.M. the curandera cleaned the mushrooms and with prayers passed them through the smoke of resin incense. While doing this she sat before a table with Christian images. She then gave the participants mushrooms in pairs, keeping thirteen pairs for herself (the sacred mushrooms are small ones and do not compare in size to *Amanita muscaria*). Wasson was given six pairs which he chewed over a period of half an hour. He described their taste as acrid and their odor as rancid.

Before midnight the curandera extinguished the light and about 30 minutes later both Wasson and Richardson started having visions which became quite intense late in the night and remained so until about 4:00 A.M. Wasson stated that the visions emerged from the center of the field of vision, were vivid in color, and always harmonious. They began with art motifs and evolved into palaces with courts, arcades, and gardens. Later he saw a mythological beast drawing a regal chariot, camel caravans, and mountains. The visions were not blurred but were sharply focused. In the meantime the curandera was singing and waving her arms rhythmically and at one time began a rhythmic dance. Both white participants fell asleep about 4:00 A.M. but woke at 6:00 A.M., rested and with clear heads but considerably shaken by their experience. Three days later Wasson repeated the experience and again had visions but not the same ones. Six weeks later in New York he ate dried mushrooms that he had collected in Mexico and found that if anything they had gained in hallucinogenic potency. In the summer of 1969, Lowy[120] also participated in a mushroom-eating ceremony in Huautla de Jiménez in Mexico. Again the ceremony was at night (beginning about 9:40 P.M.), the curandera was a woman and she offered the fresh mushrooms to the participants in pairs (five pairs to Lowy). In this instance Lowy had previously identified the sacred mushrooms as *Psilocybe caerulescans* var. *mazatecorum* (see Figure 79); he described their taste as that of radishes but more pungent and unpleasant. After the mushrooms were eaten there were no effects for about 20 minutes and then came light-headedness and a tingling sensation in the extremities followed by nausea, which caused him to vomit. The nausea passed and in about 30 minutes came a feeling of euphoria. This became intensified and both Lowy and his companions became hilarious and everything that they saw or did seemed to them to be cause for laughter. The hilarity lasted about 15 minutes and was then followed by a stage in which Lowy alternated between a feeling of contentment and expanded perception and one of acute anxiety. Among his early responses were vivid hallucinations in color and he felt as though he were inside a kaleidescope. Wasson[188] described his visions as "not blurred but in sharp focus," but Lowy stated that the patterns were amorphous rather than sharply geometric and that there was more darkness than light. He saw no objects of identifiable morphology, only "an immeasurably deep, silent, void." The ceremony attended by Lowy lasted about seven hours.

From the brief personal accounts that Wasson and Lowy gave of their experiences, it is obvious that both hallucinated, but the quality of their visions differed. They may have eaten different species of mushrooms, but it may simply reflect individual differences in reaction to the same mushroom.

Chemistry of the Hallucinogenic Mushrooms

Although a variety of chemical compounds have been isolated from *Amanita muscaria*, and such compounds as ibotenic acid, muscimol, and muscazon are generally believed to participate in the induction of the hallucinogenic syndrome that

FIGURE 79. U.S. species of *Psilocybe.* Above, *P. foenisecii;* below, *P. elongatipes.* Kauffman[106] listed *P. foenisecii* as edible; however, many Mexican and Central American species of this genus are known to be hallucinogenic. (Photographs by G. F. Atkinson.)

this mushroom is reported to produce, the picture is still far from being clear. Since the structures of these compounds have been presented earlier and reports of their actions briefly discussed, no attempt will be made here to explain that which obviously is not known at the present. However the situation with respect to the sacred mushrooms of Mexico is somewhat different.

When Wasson[188] first described his eating of the sacred mushrooms with their resultant effects, nothing was known about the chemistry of these fungi. However, this deficiency was soon corrected and in a remarkably short period of time (a manifestation of the growing interest in hallucinogens?) the responsible compounds were isolated, identified and synthesized (Hoffman et al.[98]). These substances are two closely-related indole compounds, psilocin and psilocybin (Figure 80). Since their discovery they have been subjected to a large amount of experimentation (e.g., Boszorményi,[27] Vernet,[184] Hidalgo,[97] etc.) and there is no question but that these compounds are true hallucinogens. Whether or not they may have some medical use as in the study of schizophrenia has yet to be determined. Taber[175] has reported on the fermentative production of hallucinogenic indole compounds, and Tyler,[180] on the indole derivatives of North American mushrooms. Both psilocin and psilocybin are simpler in structure than LSD and from all reports to date are not dangerous, as is LSD. The principal effects of psilocin and psilocybin are not fully understood

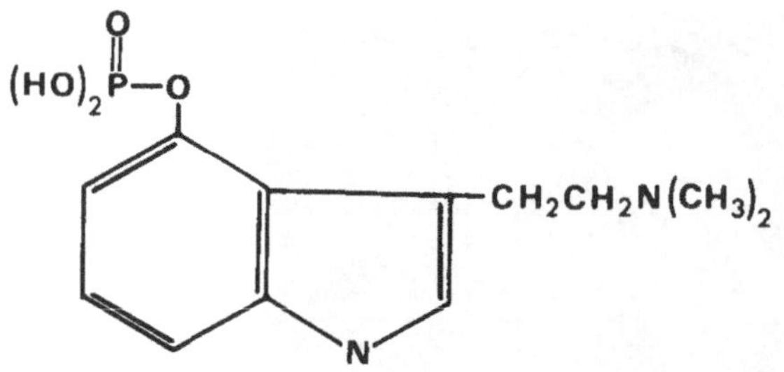

Psilocybin

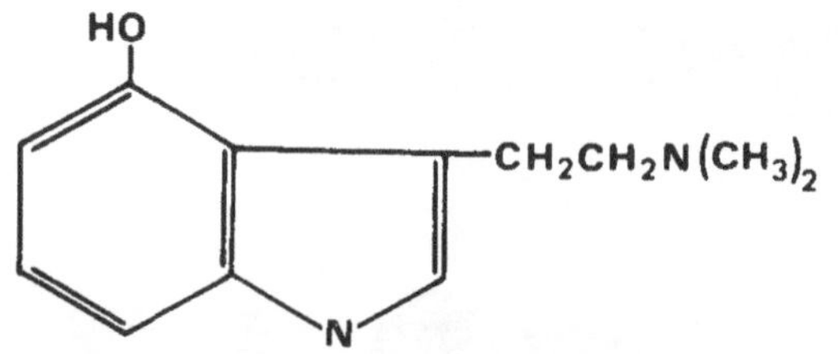

Psilocin

FIGURE 80. Psilocin and psilocybin, hallucinogenic compounds isolated from species of *Psilocybe, Conocybe, Stropharia,* and *Panaeolus.*

but they seem to occupy the serotin receptor at synaptic junctions. Being normally inhibitory, when serotin synapses are interfered with there is an excessive stimulation of cells which may lead to hallucinations.

With the present interest in hallucinogens and the knowledge that some mushrooms contain such compounds, it is entirely possible that we may learn more about the chemistry of mushrooms in the next decade than we have in the past century. Additional hallucinogenic species will undoubtedly be discovered and additional observers will report the effects of eating these mushrooms.

THE PREPARATION OF MUSHROOMS

Most people seem to have an innate tendency to fear or distrust that with which they are unfamiliar, and unfortunately to most people (and especially those in the U. S.), mushrooms are unknown quantities. Even today, in this supposedly enlightened society, it is not uncommon to observe individuals who gingerly toy with the mushrooms served with their steaks or salads and attempt to unobtrusively move them aside to avoid eating them. Some individuals simply do not care for the taste of mushrooms and this is certainly a matter in which they should have complete freedom of choice. However, questioning will sometimes reveal that the individual is literally afraid to eat mushrooms, in spite of the fact that the common cultivated mushroom has been known to be safe for eating purposes for nearly three centuries, and its production involves a multimillion dollar industry in the U. S. alone. Perhaps•after all there is in our cultural heritage a vaguely remembered taboo (possibly imposed by early Christianity) against the eating of mushrooms, which had its origin in earlier attempts to halt the eating of hallucinogenic mushrooms in connection with pagan rituals.

In spite of the fact that a great many classical writers as well as others since their time have extolled the gastronomic qualities of various mushrooms, most people in the U. S. today do not realize that a great variety of unique and tasty dishes may be prepared using mushrooms either as the basic ingredient or as a condiment to impart a desirably altered flavor. While Reitz's[149] statement that mushrooms are "the food of the gods" is somewhat extreme, the author must agree that many excellent dishes may be prepared through the proper use of such fungi. In connection with Reitz's above-quoted statement it must also be noted that if a mushroom was at one time deified by man (see Wasson[193]), it seems highly improbable that the gods would eat one of their own fellows, although among the whimsical members of the early Greek pantheon, apparently anything was possible.

It has been stated frequently that Americans of southern or central European extraction are more prone to be mushroom eaters than Nordic or Anglo-Saxon peoples; this may in general be correct but there are known instances which might cause this statement to be debated. For example, a notable exception to any generalization regarding mycophobia among Anglo-Saxon peoples is provided by one diary notation of Sir Samuel Baker, the Victorian discoverer of the Albert N'Yanza and contributer to the exploration of the upper Nile. After months of hardship, illness, and danger in the headwaters of the Nile and Great Lakes areas of Africa (although he appears to have thrived on these adventures), Baker[12] made the following entry in his journal for March 27, 1864: ". . . This river [the Kaiigiri River] rises in the great marsh that we had crossed on our way from M'rooli to Vacovia. In this neighborhood we gathered some mushrooms – the true *Agaricus campestris* of Europe – which were a great luxury." Since this opinionated Victorian gentleman usually said exactly what he thought, the above statement may be correctly interpreted to mean that at least one Englishman was a mycophile and knew enough about edible species of mushrooms to recognize one when he saw it, even in darkest Africa. Although there is presently some doubt that certain Viking groups did have the habit of eating *Amanita muscaria*, if they did it would not indicate that all Nordic peoples were always mycophobic.

Perhaps one of the most useful works for the individual who is interested in learning various ways in which mushrooms may be prepared for eating is the book of Reitz[149] entitled *Mushroom Cookery*, a book devoted entirely to mushroom recipes. In this useful volume the author has assembled nearly 150 different recipes (although she is far from having exhausted the field), all of which include mushrooms as ingredients. Other books dealing solely with mushrooms and how to cook them are those of Coffin and Lewis,[42] Faubion,[67] McKenny,[124] and the cookbook produced by the Puget Sound Mycological Society (200 Second Avenue North, Seattle, Wash. 98109).

Almost any good cookbook will include several recipes for the preparation of mushroom dishes and often truly gourmet dishes may be found described in such books. An example of this type of work is the unusual book of Dick.[53] On the other hand, the curious (and sometimes maddening) book of Herter and Herter[96] is virtually valueless to the connoisseur of mushrooms, since the principal reference to the use of mushrooms in cooking is the questionable statement that "The flavor of truffles is about the same as that of mushrooms raised here in North America commercially, in fact our mushrooms are actually much better tasting than truffles." Anyone who has compared the taste of truffles (Figure 81) with that of the common cultivated mushroom might be inclined to disagree with the Herters' statement. A very useful brochure is the *Moonlight Mushroom Recipe Book*,[8] which is available upon request from the Butler County Mushroom Farm, Inc., West Winfield, Pa. For the weight-watcher, the American Mushroom Institute has published a

FIGURE 81. Truffles, *Tuber melanospermum*. The two truffles (total weight, 12.5 g) were the sole contents of a small can purchased in Paris; cost – $1.75.

small pamphlet* entitled *Slimming Ways With Mushrooms*. For the kosher household the book *From Generation to Generation* will be found to contain several good recipes involving mushrooms (see Wissoker[202]) and the *Wild Game Cookbook* (Anonymous[7]) also contains several interesting mushroom recipes.

To give the reader a broad view of the great variety of ways in which mushrooms may be prepared, the list compiled as Table 3 is presented. In this list more than 200 named mushroom-containing dishes are grouped according to type and usage. While this may appear to be a rather long listing, it is far from being complete and the reader may add to it merely by consulting newspapers, magazines, TV guides, and cookbooks other than those listed here.

One of the common mistakes made in the preparation of mushrooms is overcooking, and most authors make the specific point that this should be avoided. In fact many individuals like the tast of raw mushrooms and when they are properly sliced, fresh uncooked mushrooms make tasty and attractive substitutes for potato chips or crackers for use with a dip. Reitz[149] also specifically stated that mushrooms should not be washed, peeled, or soaked; however, many recipe writers suggest that the mushrooms be washed first. I prefer to wash them, but they should not be soaked, just rinsed quickly under running water and blotted or placed on paper towels to dry. The commercial mushroom (*Agaricus bisporus*), which is the most widely used mushroom in the U. S., should not be peeled. However, there are several wild mushrooms of which the upper surface of the

*American Mushroom Institute, P. O. Box 373, Kennett Square, Pa. 19348

TABLE 3

Named Mushroom-containing Dishes Grouped According to Type and Usage

Hors d'oeuvres

Baked mushrooms
Basic baked stuffed mushrooms
Boiled mushrooms
Broiled mushrooms
Calf's liver pâté

Crab-stuffed mushrooms
Cream chesse mushroom spread
Dairy liver
Egg mushroom spread
French fried mushrooms

Ham rolls
Marinated mushrooms
Mushroom chips
Mushroom deviled eggs

Mushrooms in batter
Mushroom liver pâté
Mushroom pinwheels
Mushroom puffs
Mushroom toast

Mushroom turnovers
Nutty mushroom balls
Pâté of pork and mushrooms
Pickled mushrooms
Pickled mushrooms and onions

Pickled slim mushrooms
Puffball wafers
Stuffed mushrooms

Soups

Cream of mushroom soup
Cream of wheat mushroom soup
Czech mushroom soup
Gazpacho
Hungarian mushroom soup

Javanese mushroom soup
Mushroom barley soup
Mushroom barley soup with meat
Mushroom barley soup with milk
Mushroom chicken broth
Mushroom chowder

Mushroom madrilène
Mushroom matzo balls
Mushroom potato soup
Mushroom turkey soup
Mushroom vichyssoise

Mushroom watercress soup
Ragout soup
Sherried mushroom soup
Ujházy soup
Yogurt mushroom soup

First Course or Lunch

Anchovy mushrooms
Cheese and bacon mushrooms
Chicken-rice mushrooms
Curried eggs and mushrooms with cheese
Herbed mushroom omelette

Mostly mushroom salad
Mushrooms à la king
Mushrooms and eggs
Mushrooms and shrimp oriental
Mushroom and turkey club salad

Mushroom-filled pancakes
Mushroom fritters
Mushroom omelettes
Mushroom pizza

Mushroom quiche
Mushroom slimburgers
Mushroom tartlettes
Mushrooms under glass
Mushroom zucchini salad

Oysters and mushrooms
Oysters and mushrooms, creamed
Oysters and mushrooms, au gratin
Oysters on mushrooms
Scallops in shells

Seafood parmesan
Slim shrimp salad
Stuffed morels
Zingara mushrooms

TABLE 3 (continued)

Entrees

Anatolian hamburger	Mushroom mousse
Beef bourguignonne	Mushroom pie
Chicken and mushrooms in wine	Mushroom quenelles
Chicken marengo	Mushroom roast
Chicken paprikash	Mushroom seafarer's casserole
Chicken (or pheasants) under mushroom sauce	Mushroom soufflé
Filet of beef Wellington	Olive-mushroom meat loaf
Fish filets with mushrooms	Oxtail ragout
Fish steaks in mushroom sauce	Parmesan meat balls in mushroom gravy
Ham encroute with mushrooms	Pork chops with mushrooms
Ham with mushrooms	Pushcart polpetta
Meal in a pepper	Roast chicken with mushroom stuffing
Mushrooms and eggs in patty shells,	Shrimps, mushrooms, and bacon
Mushrooms and meatball kabobs	Steak and mushroom casserole
Mushroom beef Stroganoff	Steak smothered in mushrooms
Mushroom burgers fancy	Stuffed hamburgers fancy
Mushroom burgers simple	Stuffed hamburgers simple
Mushroom casserole	Tuna fish, noodles, and mushrooms
Mushroom chili	Veal chops with mushrooms
Mushroom gumbo	Veal kidneys with mushrooms
Mushroom meat loaf	Veal scallopini with mushrooms

Mushrooms as a Vegetable and with Vegetables

Bean sprouts and mushrooms	Mushrooms in sour cream
Beefy mushrooms	Mushrooms in sour cream sauce
Broiled tomatoes and mushrooms	Mushroom pudding
Cauliflower and mushrooms in cream sauce	Mushroom solianka
Collard greens	Mushroom tzimmis
Curried mushrooms	Mustard greens
Dandelion greens	Potato mushroom pancakes
Dried limas with dried mushrooms	Potato mushroom pudding
Flaming mushrooms	Princess pancakes
14 calorie mushrooms	Queen potatoes
Green beans and mushroom casserole	Sauerkraut with mushrooms
Green beans and mushroom paprika	Sautéed mushrooms
Mashed potatoes with mushrooms	Sautéed mushrooms with peas
Mushrooms à la Bordelaise	Sherried mushrooms
Mushrooms almondine	Sliced potatoes with mushrooms
Mushrooms and chestnuts	Smothered mushrooms
Mushrooms and string beans	Stewed mushrooms
Mushrooms baked in foil	Spinach greens
Mushrooms, chestnuts, and celery	Spinach and sautéed mushrooms
Mushrooms in cream	Waist-not vegetables

TABLE 3 (continued)

Salads, etc.

Christmas mushroom salad	Mushroom potato salad
Fresh mushroom dressing	Mushroom salad
Gazpacho aspic	Mushroom slaw
Marinated mushroom salad	Ritual relish
Marinated vegetable salad	Stuffed mushrooms gazpacho

Sauces

Basic mushroom sauce I	Mushroom cream sauce
Basic mushroom sauce II	Mushroom duxelles sauce
Basic mushroom sauce II with wine added	Mushroom essence
	Mushroom mayonnaise
Creamy mushroom sauce	Mushroom purée sauce
Enriched mushroom sauce	Mushroom spagetti sauce
Fine herb mushroom sauce	Red wine mushroom sauce
Madeira mushroom sauce	Supreme mushroom sauce
Magyar mushroom sauce	Truffle sauces
Mushroom cheese sauce	Velvet mushroom sauce
Mushroom cottage cheese sauce	White wine mushroom sauce

Mushroom Stuffing, Duxelles, Bread, etc.

Broiled mushroom cheese sandwiches	Mushroom bread stuffing
Duxelles butter	Mushroom Bulgar stuffing
Duxelles dip	Mushroom duxelles
Duxelles spread	Mushroom kasha stuffing
Duxelles stuffing	Mushroom potato stuffing
Easy pour pizza	Mushroom rice stuffing
Mushroom biscuits	Mushroom stuffing
Mushroom bread	Wild rice mushroom stuffing

pileus is gelatinous or slimy and many people prefer to peel such types, especially ones such as *Pholiota adiposa*, the Fat Pholiota. Most puffballs (other than the Gem-Studded Puffball) also require peeling because of their rather tough peridia.

Following is a list of recipes of several mushroom preparations selected from the eight major categories of dishes listed in Table 3. The examples presented should not necessarily be considered the best, but are presented primarily to illustrate types of ingredients needed, procedures to be followed, and above all, to give some idea of the wide range of possibilities that exists when mushrooms are used. For example, only one recipe for pickling mushrooms is included; however, there are many ways to pickle mushrooms and the resultant product may be as simple or as exotic as desired.

Hors d'oeuvres

In these days of "instant living" it is often considered necessary that such items as hors d'oeuvres be purchased ready to serve. However, insofar as I am aware, the only ready to serve items in this category which appear on the market are marinated or pickled mushrooms. Quite frequently these are the Select Boletus, *Boletus edulis* (Figure 82), imported from Poland, and in the writer's opinion this species does not lend

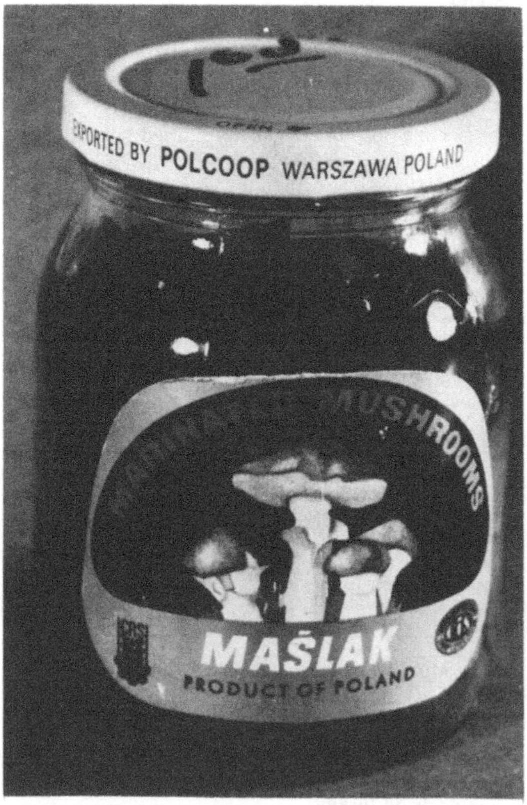

FIGURE 82. Canned marinated Select Boletus from Poland. These are available in various American markets, but this species is a poor one for marinating or pickling.

itself well to pickling. Therefore, those individuals who wish to serve tasty and interesting mushroom hors d'oeuvres must prepare them themselves.

Mushroom Pinwheels (Wissoker[202])

1 small loaf sliced bread
1 small onion, chopped
1 tsp. butter
1 pint fresh mushrooms
1/4 tsp. salt
1/4 tsp. pepper
1/4 tsp. garlic powder
1/2 tsp. MSG
1/2 tsp. Worcestershire sauce
3 oz. cream cheese
1/4 to 1/2 lb. melted butter

Cut crusts from bread. Sauté onion in 1 tsp. butter, add finely chopped fresh mushrooms and sauté together for about 5 minutes. Season mixture with above spices. Let mixture cool, then smooth cream cheese into it. Spread on bread slices, roll, and hold together with toothpicks. Can be frozen; when ready to serve, cut each roll into three parts, brush well with melted butter, and broil until lightly browned, about five minutes. Serve immediately. – Mrs. Joseph Friedman.

Puffball Wafers

fresh white-centered puffballs
eggs
cracker crumbs (fine)
salt
butter or margarine

Peel and slice fresh puffballs into 1/4 in. thick slices. If puffball is large, cut slices into pieces 1 to 2 in. in diameter (Figure 43). Dip slices in salted egg batter, then in cracker crumbs and fry in butter or margarine until golden brown (Figure 83). Serve as soon as cool enough to handle. The taste may be varied by substituting peanut oil or corn oil for butter.

Pickled Mushrooms and Onions

1/3 cup red wine vinegar
1/3 cup salad oil
1 small onion, sliced thin and separated into rings
1 tsp. salt
2 tsp. dried parsley flakes
1 tsp. prepared mustard
1 tbsp. brown sugar
2 6 oz. cans mushroom crowns, drained (about 2 cups)

Combine all ingredients except mushrooms in small saucepan and bring to boil. Add mushrooms; simmer 5 to 6 minutes. Pour into small bowl, cover, and chill several hours or overnight. Drain and serve. – Mrs. John Skok.

FIGURE 83. Puffball wafers make crisp, tasty hors d'oeuvres. They are at their best when served hot and their taste may be varied by frying in butter, peanut oil, or corn oil.

I have found that this recipe works well with small, whole basidiocarps of the Gem-Studded Puffball, *Lycoperdon gemmatum* (Figure 84). When young and very fresh the peridia of this small puffball species are tender, and if the points of attachment of the basidiocarps are trimmed off and the basidiocarps washed quickly under running water, they may be pickled whole. In season, several hundred closely packed basidiocarps of this species may be collected from a single log.

Crab-stuffed Mushrooms (Anonymous[8])

12 large mushrooms
1/4 cup salad oil
1 can (7 1/2 oz.) crabmeat
1 egg, lightly beaten
2 tbsp. mayonnaise
2 tbsp. chopped onion
1 tsp. lemon juice
3/4 cup soft bread crumbs
2 tbsp. butter or margarine, melted

Rinse and pat dry mushrooms. Remove stems (save for use in sauces, soups, etc.). Brush mushroom caps with oil. Place in a buttered baking dish. Combine crabmeat, egg, mayonnaise, onion, lemon juice, and 1/2 cup of breadcrumbs. Fill mushroom caps with mixture. Combine remaining 1/4 cup breadcrumbs with melted butter; sprinkle over crab mixture. Bake in preheated moderate oven (375° F) for 15 minutes.

Soups

Potato Mushroom Soup

1/4 cup butter or margarine
1/2 cup chopped onion
3 medium celery stalks, chopped
1 1/2 cups fresh or large handful dried mushrooms
1 tbsp. caraway seed
1/4 cup flour
3 cups water or 1 1/2 cups milk and 1 1/2 cups water
3 cups cubed raw potatoes
chopped chives or chopped tops of green onions
salt
pepper

If dried mushrooms are used, boil until tender. In large covered saucepan sauté onion, celery, caraway seed, and mushrooms in butter until tender. Gradually add some water to flour, stirring until smooth. Add flour mixture, potatoes, salt, and pepper to sautéed vegetables and chives. Cover. Bring to a boil and simmer for about 20 minutes, until potatoes are tender. Stir in 3 cups of water (or water-milk mixture). Heat thoroughly. – Adapted from Blei.[22]

Mushroom Turkey Soup (Anonymous[8])

1 lb. mushrooms
3 tbsp. butter or margarine
2 cans (10 1/2 oz. each) condensed chicken or rice soup
2 soup cans water
2 cups diced turkey
4 tsp. instant minced onion
1/4 tsp. tarragon leaves, crumbled

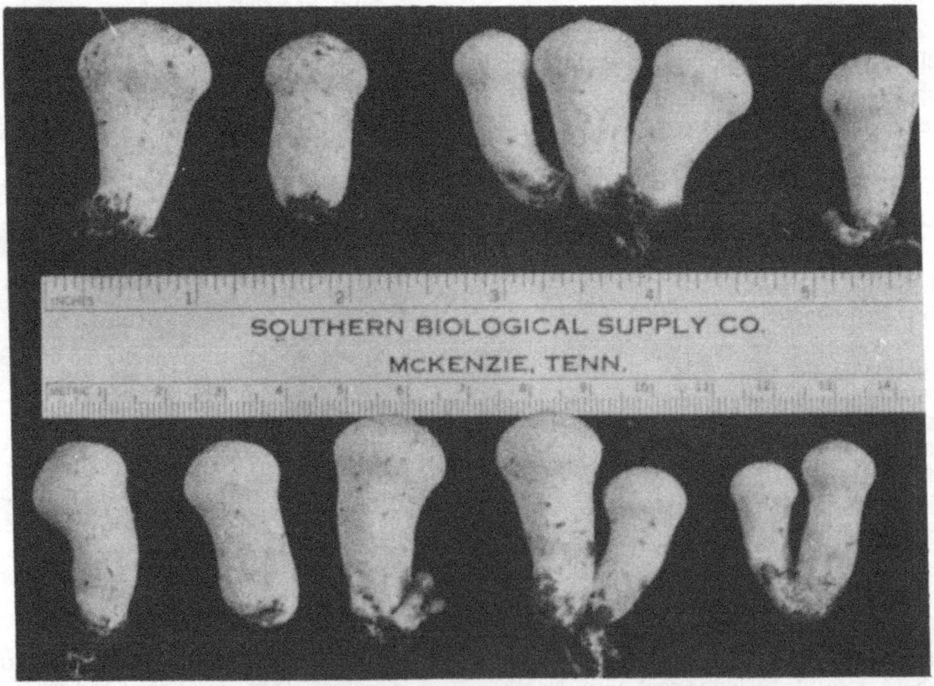

FIGURE 84. The Gem-Studded Puffball, *Lycoperdon gemmatum*. With fresh basidiocarps of this size the lower ends may be cut off and they may be pickled whole.

Rise, pat dry, and slice mushrooms (makes about 5 cups). Heat butter in a large skillet. Add mushrooms and sauté until golden, about 5 minutes. In a large saucepot combine soup, water, turkey, minced onion, and tarragon. Bring to boiling point, stirring frequently. Reduce heat and simmer 5 minutes. Stir in sautéed mushrooms. Season with salt and white pepper if desired. Serve as a main dish with French bread or tossed green salad.

First Course or Lunch

Mushrooms and Eggs

3/4 cup mushrooms
3 tbsp. butter
5 well-beaten eggs
5 tbsp. cream
1/2 tsp. salt
1/2 tsp. paprika
3/8 cup grated cheese
toast

Sauté mushrooms in butter in skillet until golden brown. Remove mushrooms and place in skillet where heat is low; add salt and paprika and blend. Add eggs beaten with cream and stir constantly from the bottom, lifting cooked eggs so that the uncooked portion runs to bottom of skillet. Cook until eggs are set. Add cheese, stirring well until melted. Add mushrooms and serve at once on hot toast. – Based on Henibuch[93] (Mrs. Donald Richard, Bethlehem Grange, Coshocton County, O.).

Stuffed Morels

large, well-shaped morels
finely chopped chicken
chopped parsley
egg yolk
powdered cloves
powdered nutmeg
flour
salt

Wash and boil morels 3 minutes, discarding the water. Dry and dust with flour. Split morels lengthwise and stuff the caps with a mixture of chicken and parsley (2 parts chicken, 1 part parsley), bound with egg yolk. Season the stuffing with salt and a little cloves and nutmeg. Butter a casserole, put in the stuffed morels, and bake in a 350° oven for 15 minutes. – Mrs. John Baxter. See Tusa.[179]

Herbed Mushroom Omelette (Anonymous[8])

1/2 lb. mushrooms
5 tbsp. butter or margarine
1 tsp. tarragon leaves, crumbled
8 eggs
1/2 tsp. onion powder
1/2 tsp. salt
dash ground black pepper

Rinse, pat dry, and slice mushrooms (makes about 2 1/4 cups). In a large skillet heat butter; add mushrooms and tarragon and sauté 4 to 5 minutes, stirring occasionally. Beat eggs with onion powder, salt, and black pepper. Pour over mushrooms in skillet. Cook over medium heat. Loosen set portion with spatula; tilt pan to let uncooked portion run underneath. Continue cooking until omelette is set but not dry. To turn out, loosen edge with spatula.

Entrees

Mushroom Pie

4 tbsp. butter
1 large onion, chopped
1 lb. fresh mushrooms or 1 lb dried mushrooms, sliced
1/4 cup flour
1 tsp. salt
1/2 tsp. pepper
1 cup half-and-half milk
1 cup chicken broth
4 tbsp. sherry

Sauté mushrooms in butter until tender. Sprinkle with flour, add salt and pepper, and coat mushrooms with flour and seasoning. Stir in half-and-half and chicken broth slowly and stir constantly until thickened. Put in 2 1/2 qt. casserole.

Place sourdough or regular biscuits on top. Bake in 350° to 375° oven for 15 to 20 minutes. – Mrs. John Baxter. See Tusa.[179]

Olive-mushroom Meat Loaf

1/2 lb. mushrooms, diced
2 tbsp. butter or margarine
1 cup minced onion (1 large onion)
1 1/2 lb. ground chuck
1/2 lb ground pork
1/2 tsp. salt
1 cup sliced stuffed olives
2 cups soft bread crumbs (4 slices bread)
2 eggs
2 tsp. salt
1/8 tsp. pepper

Preheat oven to 375°. Melt margarine or butter in skillet over medium heat; add onion and cook for 2 minutes with frequent stirring. Add mushrooms and sprinkle with salt. Cook for 1 minute and then combine mushroom-onion mixture, chuck, pork, olives, bread crumbs, eggs, salt, and pepper in a large bowl. Mix gently but thoroughly. Press mixture into a 9 x 5 x 3 in. loaf pan. Meat loaf may be baked in this pan or it may be turned out onto a greased, shallow baking pan; bake 1 hour. Makes 6 to 8 servings. – Based on Crawford.[49]

Steak and Mushroom Casserole (Dick[53])

6 beef tenderloin steaks (about 6 oz. each)
1/4 lb. butter
1 1/2 lb. fresh mushroom caps without stems
1 1/2 large Bermuda or white onions, sliced
1 1/2 cans condensed cream of mushroom soup
3/4 cup thick sour cream
3 tbsp. chopped or minced parsley

3/4 tbsp. salt
3/8 tsp. coarsely ground black pepper
3/8 tsp. dry mustard
1 1/2 cans (4 oz. each) julienne-style sliced pimentos
1/4 tbsp. thyme

Flatten steaks slightly and sauté in butter until browned. Remove to platter. Wash mushroom caps and shake off excess water but do not dry. To pan juices add mushroom caps and sliced onions. Cover pan and cook until onions are soft but not browned and liquid has been reduced to not quite dry, approximately 15 minutes. Add rest of ingredients and mix well. Correct seasoning to taste. Combine sautéed steaks with mushrooms and sauce, mix well, and put into well-buttered baking pan. This dish may be made ahead and kept in the refrigerator or freezer until needed. Bake 1 hour at 350° just before serving.

Mushrooms as a Vegetable and with Vegetables

Broiled Tomatoes and Mushrooms (based on O'Keefe[138])

2 tomatoes
1 can whole mushrooms
2 tsp. Worcestershire sauce
2 tsp. vegetable oil
2 tsp. prepared mustard with onion

Cut tomatoes in half; scoop out centers slightly and place on broiler pan. Drain mushrooms well. Combine vegetable oil, Worcestershire sauce, and mustard in small bowl and mix well; baste mushrooms in mixture. Lift mushrooms out of mustard baste with slotted spoon so they will drain and spoon into tomatoes. Broil about 3 in. from flame for 4 minutes.

Curried Mushrooms (Anonymous[8])

1/2 lb. mushrooms
4 tbsp. butter or margarine
3/4 cup minced onion
1 small clove garlic, crushed
1 tsp. curry powder
1/3 cup beef broth
1/4 tsp. salt

Rinse, pat dry, and slice mushrooms (makes about 2 1/2 cups). In a medium saucepan heat butter; add mushrooms and onion and sauté until lightly browned, about 5 minutes, stirring occasionally. Add garlic and curry powder to mushroom mixture and cook 2 minutes longer or until curry powder is darkened. Add broth and salt. Reduce heat; simmer 8 to 10 minutes. Serve over cooked rice or noodles if desired.

Green Beans and Mushroom Paprika

3 9 oz. packages frozen green beans
1/2 cup butter or margarine
3 finely chopped shallots or green onions
2 tsp. lemon juice
1 lb. mushrooms, cleaned and sliced
1 tsp. salt
1 tsp. flour
1 tsp. paprika
1/2 cup heavy cream
dash of bottled hot pepper sauce

Cook beans according to package directions and drain. Melt butter in large skillet. Sauté shallots (or green onions) over medium heat until tender and add mushrooms and lemon juice. Cook, stirring constantly, until mushrooms are tender. Combine paprika, flour, and salt and sprinkle over mushrooms. Cook with stirring for 1 minute. Add pepper sauce and cream and bring to boiling; cook for 1 minute. Add beans and blend well. Makes 8 servings. – Based on Crawford.[48]

Mushrooms and Chestnuts

1 lb. chestnuts
1 lb. mushrooms
1 1/2 cups heavy cream
3 tbsp. flour
1/2 tsp. salt
1/8 tsp. paprika
chopped parsley

Boil chestnuts until tender. Drain them well but do not mash. Sauté mushrooms and stir in flour until blended. Stir in cream slowly and season with salt and paprika. Stir in chestnuts and heat thoroughly. Serve garnished with chopped parsley. – Based on Rombauer.[153]

Spinach and Sautéed Mushrooms (Claire[39])

2 lb. fresh spinach
1/2 cup chopped fresh onions
1 tbsp. butter or margarine
1/3 cup sliced fresh mushrooms
2 tsp. fresh lemon juice
1 tbsp. butter or margarine
1 tsp. salt
3/4 tsp. sugar
1/8 tsp. ground black pepper

Wash spinach, chop or break coarsely, and set aside. Sauté onions in 1 tbsp. butter. Add chopped spinach. Cover and cook 10 minutes or only until spinach has wilted and is tender. In separate skillet sauté mushrooms in remaining 1 tbsp. butter and lemon juice. Add to spinach with salt, sugar, and black pepper. Garnish with additional sautéed sliced mushrooms if desired. Serve at once. Makes 6 servings.

Salads

Stuffed Mushroom Gazpacho

1/2 lb. medium size fresh mushrooms
1/2 cup diced tomato
2 tbsp. chopped celery
2 tbsp. chopped cucumber
1 tbsp. chopped radish
1 tbsp. chopped green pepper
1 tbsp. chopped onion
1/8 tsp. instant garlic powder
2 tbsp. bottled low-calorie garlic salad dressing

Rinse, pat dry, and remove stems from mushrooms (use stems in sauces, casseroles, etc.). Combine remaining ingredients; mix well. Spoon mixture into mushroom caps. Chill until ready to serve. Serve as salad or hors d'oeuvre. Yield: approximately 10 stuffed mushrooms; 10 calories per mushroom. – From *Slimming Ways With Mushrooms.*

Mushroom Salad (Dick[53])

3 lbs. very large mushrooms (fresh)
1/2 cup olive oil
3 1/2 cups water
juice of two lemons, strained
1 tbsp. chopped parsley
1 tbsp. celery salt
1 tsp. fennel
1/2 tsp. thyme
1 large bay leaf
12 coriander seeds
12 peppercorns

Wash mushrooms and slice from top through bottom of stems into 1/2 in. thick slices, being careful not to break off stems. In a saucepan mix rest of ingredients. Add sliced mushrooms. Bring to a boil and simmer mushrooms until they are tender. Let mushrooms cool in seasoned water, then carefully lift them out, letting some of the spices cling to them. Heap lightly in lettuce cups and serve with French dressing.

Mostly Mushroom Salad

1 lb. fresh mushrooms
1 cup diced celery
1 cup diced green pepper
2 tbsp. finely chopped onion
2 tbsp. salad or olive oil
1 tbsp. wine vinegar
2 tsp. salt
1/8 tsp. ground black pepper
2 tbsp. lemon juice

Rinse, pat dry, and slice mushrooms (makes about 5 cups). In a salad bowl place mushrooms, celery, green pepper, and onion. Combine remaining ingredients and pour over vegetables; toss gently. Serve immediately. Yield: 6 portions; 59 calories per portion. – From *Slimming Ways With Mushrooms.*

Sauces

Creamy Mushroom Sauce (Anonymous[8])

1/2 lb. mushrooms
1/2 cup butter or margarine
1/2 cup flour
1 can (10 1/2 oz.) condensed chicken or beef broth
1/2 cup light cream
1/4 tsp. salt
dash ground white pepper

Rinse, pat dry, and slice mushrooms (makes about 2 1/2 cups). In a medium saucepan heat butter and add mushrooms and sauté until golden, about 5 minutes. Blend in flour. Add broth slowly, cook and stir until smooth and thickened. Stir in cream, salt, and pepper. Cook only until hot. DO NOT BOIL. Serve over vegetables, fish, hard-cooked eggs, etc.

Mushroom Stuffing, Duxelles, Bread, Etc.

Mushroom Stuffing

1/2 lb. fresh mushrooms
1 cup shredded carrots
1 cup diced celery
3/4 cup diced onion
1/2 cup nonfat dry milk
1 tbsp. chopped parsley
1 tsp. salt
1/4 tsp. ground sage
1/8 tsp. ground marjoram
dash ground red pepper

Rinse, pat dry, and finely chop mushrooms (makes about 2 and 3/4 cups). In a medium saucepan combine mushrooms with remaining ingredients. Place over low heat and cook, stirring, until liquid appears in pan, about 5 minutes. Cover and continue to simmer 30 minutes longer, stirring occasionally. Serve hot as is or place in cavity of small chicken before roasting. This recipe may be doubled or tripled. – From *Slimming Ways with Mushrooms.*

Mushroom Biscuits (Anonymous[8])

1/2 lb. mushrooms
2 tbsp. butter or margarine
1/4 cup finely chopped onion
1/4 tsp. ground thyme
1/16 tsp. ground white pepper
2 cups biscuit mix
2/3 cup milk

Rinse, pat dry, and coarsely chop mushrooms (makes about 3 cups). In a medium skillet heat butter and add mushrooms and onion; sauté until lightly browned, about 5 minutes, stirring occasionally. Sprinkle with thyme and white pepper. Cool slightly. Blend mushroom mixture into biscuit mix. Add milk and stir until well mixed. Drop by the tablespoonful onto greased cookie sheets 2 in. apart. Bake in a preheated hot oven (450° F) 12 minutes or until nicely browned. Serve with chicken, meat, or fish.

Mushroom-filled Pancakes

This is a Hungarian dish that may be served for lunch or supper. Hungarian pancakes are larger (8 in. in diameter) and thinner than the American variety and almost transparent. They are commonly stacked with alternating layers of filling, baked, and cut like cake. In this instance the filling is largely mushrooms.

plain pancakes
1 1/2 cups finely chopped mushrooms
2 tbsp. butter
1/2 tsp. salt
dash of pepper
1/2 cup sour cream
1 egg, slightly beaten

To prepare the filling, brown mushrooms in butter; cool slightly, then combine with salt, pepper, sour cream, and egg.

Place a freshly made pancake in a buttered baking dish and spread with 1 tbsp. of filling. Repeat until there are many layers. Butter the top cake and bake in a moderate oven (350° F) for about 20 minutes. Slice and serve hot. – Based on Bennett and Clark.[17]

Summation

From the above rather meager collection of recipes and the long list presented in Table 3, it is obvious that mushrooms may be utilized in the preparation of many tasty and different dishes. In fact, diet monotony can almost be prevented through the use of mushrooms alone. While nearly all of these recipes are based upon the use of the common cultivated mushroom, *Agaricus bisporus*, the individual who finds these and other dishes to his liking will find in many instances that certain edible wild mushrooms can be substituted in these recipes to advantage. For example, in the early spring when Morels are generally available, these mushrooms can be substituted for the commercial mushroom, yielding, in many instances, a far more delectable dish.

If one becomes acquainted with a dozen or more edible wild species of mushrooms he will soon find that their tastes are subtly different and that their cooking characteristics may be markedly different. In this writer's opinion the Oyster Mushroom, *Pleurotus ostreatus*, is a better species to use for soups than is the cultivated mushroom; however, it does not lend itself as well to pickling because it is much larger than the cultivated type (and hence must be cut into smaller pieces which do not have the eye appeal) and also does not have the firm texture of *Agaricus bisporus* buttons. The Shaggy Mane, *Coprinus comatus* (Figures 4 and 30), is an extremely tasty species but because of its very thin, closely packed gills and relatively thin flesh above the gills it is quite fragile. This species is probably at its best when carefully sautéed in butter. For more extensive information on the preparation of mushrooms for the table and a much more extensive collection of recipes, the reader is advised to consult the books and papers recommended earlier in this section.

Not infrequently it is the experience of the mushroom collector to find a much larger crop of basidiocarps of a desirable species than he can possibly eat in any reasonable period of time. Therefore, it should be noted that mushrooms may also be canned, dried, or frozen. No general directions for such processing may be given for all species of edible wild mushrooms, but in my experience only species of rather firm texture should be canned, dried, or frozen. Thus, fragile, rapidly maturing species such as the Shaggy Mane cannot be preserved for future use by these means, while the Sponge Mushroom, *Morchella esculenta*, may be preserved in all three ways, although to me, reconstituted dried Morels are more appealing than either canned or frozen ones. If the collector is fortunate enough to find great numbers of ascocarps of this species, he will find that they can be dried quite easily. First, cut off the basal parts of the stipes with adhering soil particles and then string the ascocarps like beads on stout thread or string by passing the thread (with a needle) through the lower parts of the stipes. The strings may then be hung in a dry place until the ascocarps are dry.

Although most of the mushroom recipes encountered in the American literature are quite naturally based upon the use of the common cultivated mushroom, since this species may be purchased fresh or canned the year around, the mycophagist need not confine himself to the use of this one type even though he may not be interested in collecting wild mushrooms. Canned truffles (Figure 50), canned or dried Morels, and pickled or marinated Select Boletus (Figure 82) are now available in many U. S. markets, although they are usually rather high-priced. Also available in many markets are the Padi Straw Mushroom either canned (Figure 85) or dried, dried Ear Fungus and Shiitake (see Gray[80,81]), all imported from Asia. If the mycophagist becomes a devoted one and learns to collect and identify a number of

FIGURE 85. Canned Padi Straw Mushrooms which are grown and packed in Taiwan and Hong Kong are now offered for sale in the U.S.

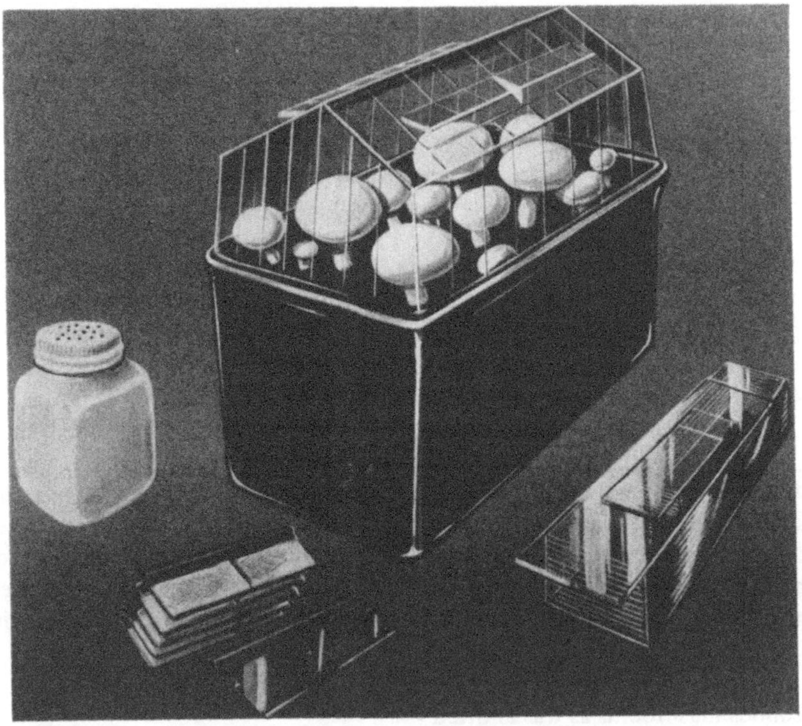

FIGURE 86. Mushroom growing kit. Small kits such as this may be used to grow mushrooms in the home or laboratory. (Photograph courtesy of Lab-Aids, Inc.)

edible wild mushrooms, he should by all means consult the cookbook produced by the Puget Sound Mycological Society, which is devoted to the preparation of wild mushrooms. One word about the purchase of fresh, cultivated mushrooms: mostly they are packaged as nearly white buttons, but an occasional package of larger, darker mushrooms with expanded pilei may be found (often at a much reduced price as compared with the buttons). It is recommended that these be purchased if to be used with steaks, roast, stuffing, vegetables or in soups or omelettes because of their much better flavor. For salads, pickled mushrooms or for use with a dip, buttons have considerably more eye appeal. At the risk of offending administration economists it is recommended that in the purchase of canned common cultivated mushrooms, the prices of those packed in Taiwan should be compared with those packed domestically; the Asian packed mushrooms are commonly much less expensive, and it is highly improbable that their purchase will measurably affect the balance of payments!

For the individual who may be interested in watching mushrooms grow and develop there is available from Lab-Aids, Inc. a mushroom growing kit (Figure 86) with which this can be accomplished.

SUMMARY

While the first part *(CRC Crit. Rev. Food Technol.,* 1(2), 225, 1970*) of this series was designed to present an overall view of the many ways in which fungi are involved in a beneficial manner in the area concerned with man's food, the section devoted to the direct use of fungi as food was largely concerned with a discussion of the few mushrooms that are deliberately cultivated by man, and only passing reference was made to the possibility of man taking more advantage of wild mushrooms as a source of food, Therefore, this second part has been designed specifically to emphasize this possibility. It is primarily aimed at the individual who enjoys eating mushrooms but is not satisfied with an occasional mushroom-smothered steak, a mushroom salad, or a bowl of creamed mushroom soup.

For the most part the edible wild mushrooms selected for discussion in the first section are common ones which the beginner can quickly learn to identify beyond question with a minimum

*This article was reprinted in hard cover edition by CRC Press under the title *The Use of Fungi as Food and in Food Processing,* 1970.

of effort. They have been described and illustrated, but the reader is again admonished that this section is meant to supplement rather than to supplant a reliable mushroom manual written by one of the experts in the field.

The section on Fungi as Animal Food was included for two reasons. First, evidence accumulated to date indicates that wild mushrooms may play a very important role in the feeding of certain edible wildlife species and thus in some instances are involved near the beginning of a food chain that ultimately ends with man. Second, the very important point is made that even though some wild animal may eat a mushroom with no apparent ill effects, this does not necessarily indicate that man can do likewise. On the contrary, it was pointed out that some mushrooms that are known to be deadly poisonous to man are eaten by animals, apparently with no untoward results as far as the animal is concerned.

Almost as much space has been given to a discussion of poisonous mushrooms as was allotted to a discussion of edible ones. Although there are many less poisonous ones than there are non-poisonous mushrooms, the collector for the table must never rely on such seemingly favorable statistics. To be quite safe the beginning mycophagist should first become acquainted with the poisonous species, especially the deadly Amanitas and the False Morel, *Gyromitra esculenta,* species which probably have caused the greatest number of fatalities due to mushroom poisoning in the U. S. It has been suggested that Ford's 1923 classification of types of mushroom poisoning be reduced from five to three categories, since his "mycetismus gastrointestinalis" may be largely instances of food allergies and his "mycetismus cerebralis" is primarily hallucinogenesis.

Hallucinogenic mushrooms were included in the present discussion because, while their use is not primarily a food-related one, they are ingested by man. A discussion of these fungi has also been used to introduce the reader to the new field of mycology known as ethnomycology. This new branch of the field has developed only during the past 15 years and already its findings are beginning to reveal that man's cultural and religious past may have involved mushrooms to a greater extent than was formerly suspected.

A brief discussion of the ways in which mushrooms may be prepared for eating was included because apparently the average individual is almost totally unaware of the many and varied ways in which mushrooms may be prepared for the table. Recipes for several dishes in each of the various categories of mushroom-containing dishes have been presented in some detail. However, no attempt was made to exhaust the subject and several cookbooks dealing specifically with the preparation of mushrooms were cited for use by the interested reader.

Many years ago I stated that in any area of man's affairs where fungi are involved (and this includes most of the important ones), if the beneficial actions of fungi are weighed against the harmful ones, the beneficial actions are in the majority. From the material presented in this work as well as in the first volume of this series, the area involving man's food does not seem to be an exception to this general rule.

Acknowledgments

In all fairness to the many individuals who have rendered assistance and encouragement, this work should probably be considered to have many silent co-authors, although they certainly should not be blamed for any mistakes or possibly unpopular opinions that it may contain. As usual, requests for assistance have been promptly and generously answered by colleagues both here and abroad. My thanks are due Dr. C. M. Christensen, Dr. B. Lowy, Dr. D. P. Rogers, and Dr. A. H. Smith, as well as the Canada Department of Agriculture for supplying various photographs. I am indebted also to Mr. R. G. Wasson, who gave me his permission to reproduce illustrations from his book, *Soma, Divine Mushroom of Immortality,* and his *Life* Magazine article. Special thanks are extended to Dr. R. P. Korf, who called my attention to and made available the fine Atkinson illustrations in the Department of Plant Pathology, Cornell University. Locally, Miss Jane Glaser and Dr. L. Hanzely provided help with some of the photography, and finally I am indebted to my patient colleagues in the Department of Biology, Northern Illinois University, who by this time must be very tired of hearing about the fungi!

REFERENCES

1. **Abul-Haj, S. K., Ewald, R. A., and Kazyak, L.,** Fatal mushroom poisoning. Report of a case confirmed by toxicologic analysis of tissue, *N. Engl. J. Med.*, 269(5), 223, 1963.
2. **Ainsworth, G. C. and Bisby, G. R.,** *Dictionary of the Fungi,* 5th ed., Commonwealth Mycological Institute, London, 1961.
3. **Alkim, U. B.,** *Archeologia Mundi. Anatolia I,* World Publishing Co., New York, 1968.
4. **Allegro, J. M.,** *The Sacred Mushroom and the Cross,* Doubleday, Garden City, N.Y., 1970.
5. **Anonymous,** *Boletus edulis* is a mushroom, *Mushroom Growers Assoc. Bull.,* 114, 208, 1959.
6. **Anonymous,** "Mushroom" soup cleared. Court decision on fungus, *London Daily Telegraph,* May 8, 1959.
7. **Anonymous,** *Wild Game Cookbook,* Jackson Hole Art Assoc., Jackson, Wyoming, 1969.
8. **Anonymous,** *Moonlight Mushroom Recipe Book,* Butler County Mushroom Farm, Inc., West Winfield, Pa., 1971.
9. **Atkinson, G. F.,** *Mushrooms, Edible, Poisonous, Etc.,* 2nd ed., Hafner Publishing Co., New York, 1961.
10. **Azar, A., Thomas, A. A., and Shillito, F. H.,** Pyridoxine and phenobarbital as treatment for aerozine-50 toxicity, *Aerosp. Med.,* 41(1), 1, 1970.
11. **Back, K. C. and Pinkerton, M. K.,** Toxicology and pathology of repeated doses of monomethylhydrazine in monkeys, AMRL-TR-66-199, *Aerosp. Med. Res. Lab.,* Wright Patterson Air Force Base, Ohio, 1967.
12. **Baker, S. W.,** *The Albert N'Yanza. Great Basin of the Nile and Exploration of the Nile Sources,* Vol. II, Sidgwick and Jackson, London, 1962 (first published in 1866).
13. **Benedict, R. G. and Brady, L. R.,** Fermentative production of toxic cyclopeptides of *Galerina marginata, Lloydia,* 30(4), 372, 1967.
14. **Benedict, R. G. and Brady, L. R.,** Further studies on fermentative production of toxic cyclopeptides by *Galerina marginata* (Fr.) Kühn, *Lloydia,* 30, 379, 1967.
15. **Benedict, R. G., Brady, L. R., Stuntz, D. E., and Spurr, J.,** Occurrence of the deadly *Amanita verna* in the Pacific Northwest, *Mycologia,* 62, 597, 1970.
16. **Benedict, R. G., Tyler, V. E., Jr., Brady, L. R., and Weber, E. J.,** Fermentative production of amanita toxins by a strain of *Galerina marginata, J. Bacteriol.,* 91, 1380, 1966.
17. **Bennett, P. P. and Clark, V. R.,** *The Art of Hungarian Cooking,* Doubleday, Garden City, N.Y., 1954.
18. **Bergoz, R. and Righetti, A.,** Intolérance aux champignons par malabsorption sélective du tréhalose, *Schweiz. Med. Wochenschr.,* 100, 1244, 1970.
19. **Bessey, E. A.,** *Morphology and Taxonomy of Fungi,* Blakiston, Philadelphia, Pa., 1950.
20. **Birch, C. A.,** Mushroom poisoning, *Practitioner,* 157, 135, 1946.
21. **Blei, N.,** On the trail of the houby, *Chicago Tribune Magazine Section,* 32, March 5, 1972.
22. **Blei, N.,** Eat your soup, *Chicago Tribune Magazine Section,* 55, March 12, 1972.
23. **Block, S. S., Stephens, R. L., and Barrets, A.,** Chemical identification of amanita toxin in mushrooms, *Science,* 121, 505, 1955.
24. **Block, S. S., Stephens, R. L., and Murrill, W. A.,** The *Amanita* toxins in mushrooms, *Agr. Food Chem.,* 3, 584, 1955.
25. **Boehm, R. and Külz, E.,** Ueber den giften Bestandtheil der essbaren Morchel (*Helvella esculenta*), *Arch. Exp. Pathol. Pharmakol.,* 19, 403, 1885.
26. **Borhegyi, S. F. de,** Miniature mushroom stones from Guatemala, *Am. Antiquity,* 26(4), 498, 1961.
27. **Boszorményi, Z.,** Psilocybin and diethyltryptamines: Two tryptamine hallucinogens, *Neuro-Psychopharmacol.,* 2, 226, 1960.
28. **Brekhman, I. I. and Sem, Y. A.,** Ethnopharmacological investigation of some psychoactive drugs used in Siberian and far-eastern nationalities of U.S.S.R. (summary), *U.S. Public Health Serv. Publ.,* 1645, 415, 1967.
29. **Buck, W. F.,** Mushroom poisoning since 1924 in the United States, *Mycologia,* 53, 537, 1961.
30. **Buller, A. H. R.,** The red squirrel of North America as a mycophagist, *Br. Mycol. Soc. Trans.,* 6, 355, 1920.
31. **Buller, A. H. R.,** *Researches on Fungi,* Vol. II, Longmans, Green & Co., London, 1922.
32. **Cann, H. M. and Verhulst, H. L.,** Mushroom poisoning, *J. Dis. Child.,* 101, 128, 1961.
33. **Catalfomo, P. and Eugster, C. H.,** Muscarine and muscarine isomers in selected *Inocybe* species, *Helv. Chim. Acta,* 53(4), 848, 1970.
34. **Catalfomo, P. and Eugster, C. H.,** *Amanita muscaria:* present understanding of its chemistry, *Bull. Narc.,* 22(4), 33, 1970.
35. **Cheney, W. F.,** Cases of mushroom poisoning, *Pac. Med. Surg.,* 5, 119, 1871-72.
36. **Child, G. P.,** The inability of Coprini to sensitise man to ethyl alcohol, *Mycologia,* 8, 200, 1952.
37. **Christensen, C. M.,** *Common Fleshy Fungi,* Burgess Publishing Co., Minneapolis, Minn., 1965.
38. **Christensen, C. M.,** Edible wild mushrooms, Extension Bull. 357, Agr. Ext. Service, University of Minnesota, 1968.
39. **Claire, A.,** Try spinach and sauteed mushrooms, *DeKalb (Ill.) Daily Chronicle,* 5, January 4, 1972.
40. **Clark, D. A., Bairrington, J. D., Bitter, H. L., Coe, F. L., Medina, M. A., Merritt, J. H., and Scott, W. N.,** Pharmacology and toxicology of propellant hydrazines, *Aeromed. Rev.,* No. 11-68, USAF School of Aerospace Medicine, Brooks Air Force Base, Texas, 1968.

41. **Cleland, J. B.**, Plants, including fungi, poisonous or otherwise injurious to man in Australia, *Med. J. Aust.*, 2, 161, 1943.
42. **Coffin, G. and Lewis, M.**, *Twenty Common Mushrooms and How to Cook Them*, International Pocket Library, Bruce Humphries, Inc., Boston, Mass., 1965.
43. **Coker, W. C.**, *The Clavarias of the United States and Canada*, University of North Carolina Press, Chapel Hill, N.C., 1923.
44. **Coker, W. C. and Beers, A. H.**, *The Boletaceae of North Carolina*, University of North Carolina Press, Chapel Hill, N.C., 1943.
45. **Coker, W. C. and Couch, J. N.**, *The Gastromycetes of the Eastern United States and Canada*, University of North Carolina Press, Chapel Hill, N.C., 1928.
46. **Coville, F. V.**, Observations on recent cases of mushroom poisoning in the District of Columbia, Circ. 13, Div. Botany, U.S. Dept. Agr., 1898.
47. **Cram, W. E.**, The red squirrel, *J. Mammal.*, 5, 37, 1924.
48. **Crawford, F. M.**, The bountiful holiday feast, *Am. Home*, 74(12), 58, 1971.
49. **Crawford, F. M.**, Old-time meat loaf was never like this, *Am. Home*, 75(3), 91, 1972.
50. **Cullinan, E. R. and Henry, D.**, Poisonous mushrooms in the Nairobi district, *East Afr. Med. J.*, 22, 252, 1945.
51. **Dabski, H., Drozd, T., and Bryc, R.**, Leczenie zatruc muchmoren sroomtnikowym kwasem tiokanowyn, *Pol. Tyg. Lek.*, 25(9), 338, 1970 (cf. Simons, 1971).
52. **Dearness, J.**, The personal factor in mushroom poisoning, *Mycologia*, 3, 75, 1911.
53. **Dick, E. B.**, *The Old House Holiday & Party Cookbook*, Cowles Book Co., Inc., New York, 1969.
54. **Donner, K.**, Ethnological notes about the Yenisei-Ostyak, *Mem. Soc. Finno-ougrienne*, 66, 81, 1933.
55. **Dreisbach, R. H.**, *Handbook of Poisoning: Diagnosis & Treatment*, 5th ed., Lange Medical Publications, Los Altos, Calif., 1966.
56. **Dubash, J. and Teare, D.**, Poisoning by *Amanita phalloides*, *Br. Med. J.*, 1, 45, 1946.
57. **Dudderar, G. R.**, A survey of the food habits of the gray squirrel (*Sciuries carolinensis*) in Montgomery County, Virginia, M.S. Thesis, Virginia Polytechnic Institute, Blacksburg, Va., 1967.
58. **Duggar, B. M.**, The principles of mushroom growing and mushroom spawn making, U.S. Govt. Print. Off., Washington, D.C., 1920.
59. **Dujarric de la Rivière, R.**, Essai de sérothérapie, contre les empoisonnements par champignons vénéneux, *Bull. Acad. Med. (Paris)*, 94, 1000, 1925.
60. **Dujarric de la Rivière, R.**, Etude d'une toxine végétale: la toxine phallinique, *Ann. Inst. Pasteur*, 43, 961, 1929.
61. **Emmons, C. W.**, Mycology and medicine, *Mycologia*, 53, 1, 1961.
62. **Emmons, C. W.**, Deaths from ingestion of mushrooms, *Mycologia*, 54, 115, 1962.
63. **Eugster, C. H.**, The chemistry of muscarine, in *Advances in Organic Chemistry*, Vol. II, Interscience, New York, 1960.
64. **Eugster, C. H.**, Isolation, structure and synthesis of central-active compounds from *Amanita muscaria* (L. ex Fr.) Hooker, *U.S. Public Health Serv. Publ.*, 1645, 416, 1967.
65. **Eugster, C. H.**, Chemie der Wirkstoffe aus dem Fliegenpilz (*Amanita muscaria*), *Fortschr. Chem. Org. Naturst.*, 27, 261, 1969.
66. **Eugster, C. H. and Waser, P. G.**, Zur Kenntnis des Muscarins, *Experientia*, 10, 298, 1954.
67. **Faubion, N. L.**, *Some Edible Mushrooms and How to Cook Them*, University of Washington Press, Seattle, Wash., 1938.
68. **Fischer, O. E.**, Mushroom poisoning, in *The Agaricaceae of Michigan, Mich. Geol. Biol. Survey*, Publ. 16, Biol. Ser. 5, 823, 1918 (reprinted by Johnson Reprint Corp., New York, 1965).
69. **Fiume, L. and Wieland, T.**, Amanitins: chemistry and action, *Fed. Eur. Biochem. Soc. Lett.*, 8(1), 1, 1970.
70. **Ford, W. W.**, The distribution of haemolysins, agglutinins and poisons in fungi, especially the Amanitas, the Entolomas, the Lactarius and the Inocybes, *J. Pharmacol. Exp. Ther.*, 2, 285, 1911.
71. **Ford, W. W.**, A new classification of mycetismus (mushroom poisoning), *Trans. Assoc. Am. Physicians*, 38, 225, 1923.
72. **Ford, W. W.**, Poisonous mushrooms, in *Legal Medicine and Toxicology*, Vol. 2, Saunders, Philadelphia, Pa., 1923, 817.
73. **Ford, W. W. and Clark, E. D.**, A consideration of the properties of poisonous fungi, *Mycologia*, 6, 167, 1914.
74. **Ford, W. W. and Sherrick, J. L.**, Further observations on fungi, particularly *Clitocybe sudorifica* Peck, *Pholiota autumnalis* Peck and *Inocybe decipiens* Bresadola, *J. Pharmacol. Exp. Ther.*, 4, 321, 1913.
75. **Franke, S., Freimuth, U., and List, P. H.**, Über die Giftigkeit der Frühjahrslorchel *Gyromitra (Helvella) esculenta* Fr., *Arch. Toxikol.*, 22, 293, 1967.
76. **Fraser, P. J.**, Pharmacological actions of pure muscarine chloride, *Br. J. Pharmacol.*, 12, 47, 1957.
77. **Genest, K., Hughes, D. W., and Rice, W. B.**, Muscarine in *Clitocybe* species, *J. Pharm. Sci.*, 57(2), 331, 1968.
78. **Gray, W. D.**, Physiological variation in isolates of *Polyporus schweinitzii*, *Trans. Ky. Acad. Sci.*, 12, 59, 1947.
79. **Gray, W. D.**, *The Relation of Fungi to Human Affairs*, Holt, New York, 1959.
80. **Gray, W. D.**, The use of fungi as food and in food processing, *CRC Crit. Rev. Food Technol.*, 1(2), 225, 1970.
81. **Gray, W. D.**, *The Use of Fungi as Food and in Food Processing*, CRC Press, Cleveland, Ohio, 1970.

82. **Grossman, C. M. and Malbin, B.,** Mushroom poisoning: a review of the literature and report of two cases caused by a previously undescribed species, *Ann. Intern. Med.*, 40, 249, 1954.
83. **Groves, J. W.,** Mushroom collecting for beginners, *Can. Dept. Agr.*, Publication 861, 1958.
84. **Groves, J. W.,** Poisoning by morels when taken with alcohol, *Mycologia*, 56, 779, 1964.
85. **Guba, E. F.,** Selected list of fleshy fungi credited with significant poisonings, Prepared by Boston Mycological Club (Bull. No. 3), June, 1961.
86. **Harshberger, J. W.,** An ancient Roman toadstool carved in stone, *Mycologia*, 21, 143, 1929.
87. **Hatt, R. T.,** The red squirrel: its life history and habits, with special reference to the Adirondacks of New York and Harvard Forest, *Roosevelt Wild Life Annals*, Vol. 2, 1, 103, 1929.
88. **Heim, R.,** Les champignons divinatoires utilises dans des Indiens Mazateques, recuilles au cours de leur premier voyage au Mexique, en 1953, par Mme. Valentina Pavlovna Wasson et M. R. Gordon Wasson, *C. R. Acad. Sci.*, 242, 965, 1956.
89. **Heim, R.,** Les champignons toxique et hallucinogènes, Paris, 1963.
90. **Heim, R.,** L'Amanite tue-mouche nord-américaine n'est pas la *Muscaria, Rev. Mycol.*, 30(4), 294, 1966.
91. **Heim, R. and Wasson, R. G.,** The "mushroom madness" of the Kuma, *Botanical Museum Leaflets,* Harvard University, 21(1), 1, 1965.
92. **Hendricks, H. V.,** Poisoning by the false morel (*Gyromitra esculenta*), *J.A.M.A.*, 114(17), 1625, 1940.
93. **Henibuch, W.,** *Ohio State Grange Cookbook,* compiled by Ohio State Grange Home Economics Committee and available through the Grange, 1952.
94. **Henrici, A. T.,** *Molds, Yeasts and Actinomycetes,* Wiley, New York, 1945.
95. **Herrick, J. A.,** Mushrooms – to eat or not to eat, *School Sci. Math,* 679, December, 1948.
96. **Herter, G. L. and Herter, B. E.,** *Bull Cook and Authentic Historical Recipes and Practices,* Herter's Inc., Waseca, Minn., 1970.
97. **Hidalgo, T. W.,** Estudio comparativo psicofisológico de la mescalina, dietalamida de ácido D-lisérgico y psilocybina, *Acta Med. Venez.*, 8, 56, 1960.
98. **Hoffman, A., Frey, A., Ott, H., Petrzikla, T., and Troxler, F.,** Konstitution-saufklärung und Synthese von Psilocybin, *Experientia,* 14, 397, 1958.
99. **Ingalls, D. H. H.,** Soma, Divine Mushroom of Immortality (book review), *New York Times Book Review,* 15, September 5, 1971.
100. **Ingalls, D. H. H.,** Remarks on Mr. Wasson's Soma, *J. Am. Oriental Soc.*, 91(2), 188, 1971.
101. **Jacobsen, T. and Richardson, C. C.,** Mr. Allegro among the mushrooms, *Union Seminary Quart. Rev.*, 26(3), 235, 1971.
102. **Jenkins, A. E.,** Old tombstone inscription recording death of a family from mushroom poisoning, *Mycologia,* 52, 521, 1960.
103. **Johnson, J. B.,** The elements of mazatec witchcraft, *Gothenburg Ethnological Museum, Ethnographical Studies,* 9, 119, 1939.
104. **Josserand, M.,** The ability of Coprini to sensitize man to ethyl alcohol, *Mycologia,* 44, 829, 1952.
105. **Kambly, P. E. and Lee, R. E.,** The Gastromycetes of Iowa, *Univ. of Iowa Stud. Natur. Hist.*, N. S., 17(4), 121, 1936.
106. **Kauffman, C. H.,** *The Agaricaceae of Michigan,* Vol. I, Mich. Geol. and Biol. Survey, Publ. 26, Biol. Ser. 5, 1918 (reprinted by Johnson Reprint Corp., New York, 1965).
107. **Kidder, A. V., Jennings, J. D., and Shook, E. M.,** Excavations at Kaminaljuyú, Guatemala, *Carnegie Inst. Wash. Publ.*, 561, 1946.
108. **Korf, R. P. and Burdsall, H. H.,** Common mushrooms: edible and poisonous, New York State Coll. Agr., Cornell University Extension Bull. 386, 1971.
109. **Korschgen, L. J.,** A general summary of the foods of Missouri's game and predatory animals, *Missouri Conserv. Comm.* (P-R 13-R-4), 1952.
110. **Krieger, L. C. C.,** *The Mushroom Handbook,* Dover Publications, Inc., New York, 1967.
111. **Kubicka, J.,** Prevention and therapy of intoxications caused by *Amanita phalloides* in South Bohemia region, *Prakt. Lek.*, 44(18), 702, 1964 (cf. Simons, 1971).
112. **Lange, M. and Hora, F. B.,** *A Guide to Mushrooms and Toadstools,* Dutton, New York, 1963.
113. **Leete, E.,** in *Biogenesis of Natural Products,* Bernfeld, P., Ed., MacMillan, New York, 1963.
114. **Lewes, D.,** Mushroom poisoning due to *Amanita phalloides, Br. Med. J.*, 2, 383, 1948.
115. **Liener, I. E.,** Toxic substances associated with seed proteins, in *World Protein Resources,* Adv. Chem. Ser., 57, 178, 1966.
116. **Limousin, H. and Petit, G.,** Essai thérapeutique au cours des intoxications par l'amanita phalloide, *Bull. Acad. Med.* (Paris), 107, 698, 1932 (translated into English by F. R. Seymour, *Br. Med. J.*, 2, 220, 1932).
117. **List, P. H. and Luft, P.,** Gyromitrin, das Gift der Frühjahrslorchel, *Arch. Pharm.*, 301(4), 294, 1968.
118. **List, P. H. and Luft, P.,** Nachweis und Gehaltsbestimmung von Gyromitrin im frischen Lorcheln, *Arch. Pharm.*, 303(2), 143, 1969.
119. **Lowe, J. L.,** *The Polyporaceae of New York State,* The N.Y. State College of Forestry, Syracuse, 1942.
120. **Lowy, B.,** Some observations on ethnomycology in Mexico and Guatemala, *Rev. Interamericana Rev.*, 1, 39, 1971.
121. **Lowy, B.,** New records of mushroom stones from Guatemala, *Mycologia,* 63, 983, 1971.

122. Martin, G. W., The numbers of fungi, *Proc. Iowa Acad. Sci.*, 58, 175, 1951.
123. McIlvaine, C., *One Thousand American Fungi,* Bowen-Merrill Co., Indianapolis, 1900.
124. McKenny, M., *The Savory Wild Mushroom,* University of Washington Press, Seattle, Wash., 1962.
125. Melendro, J. C., Las intoxicaciones mortales por setas (sindrome phalloide), *Rev. Clín. Españ.*, 17, 211, 1945.
126. Mendoza, J. M. and Leus-Palo, S., New species of amanita, *Philippine J. Sci.*, 63, 281, 1937.
127. Metcalf, M. M., *Amanita muscaria* in Maine, *Science,* N.S., 61, 567, 1925.
128. Miller, H. A. and Halls, L. K., Fleshy fungi commonly eaten by southern wildlife, Southern Forest Exp. Sta., New Orleans, USDA Forest Ser. Res. Paper, SO-49, 1969.
129. Mittman, W., Zur Klinik und Therapie der Lorchelvergiftung (*Gyromitra esculenta*), *Z. Aerztl. Fortbild.*, 67, 710, 1968.
130. Miura, O., *Clitocybe acromelalgia* Ichimura, and disease resembling erythromelalgia which it causes, *Tohoku J. Exp. Med.*, 31, 1, 1937.
131. Mullendore, B., Michigan's morels, *Mich. Nat. Res.*, 40(3), 15, 1971.
132. Munilla, A., Rosa-Mato, F., and Buno, W., Toxic fungi in Uruguay, *Arch. Soc. Biol. Montevideo,* 6, 141, 1935.
133. Murrill, W. A., Poisonous mushrooms, *Mycologia,* 2, 255, 1910.
134. Murrill, W. A., Illustrations of fungi – XXV, *Mycologia,* 8, 231, 1916.
135. Myler, R. K., Lee, J. C., and Hopper, J., Renal tubular necrosis caused by mushroom poisoning, *Arch. Intern. Med.*, 114, 196, 1964.
136. Nixon, C. M., Worley, D. M., and McLain, M. W., Food habits of squirrels in southeast Ohio, *Ohio J. Wildlife Manage.*, 32, 294, 1968.
137. O'Flaherty, W. D., The post-Vedic history of the Soma plant, in *Soma, Divine Mushroom of Immortality*, Wasson, R. G., Ed., 95, 1969.
138. O'Keefe, J., Mission possible: meals in 31 minutes, *Family Circle*, 80(2), 95, 1972.
139. Overholts, L. O., *The Polyporaceae of the United States, Alaska and Canada*, University of Michigan Press, Ann Arbor, 1953.
140. Patrick, R. L. and Back, K. C., Pathology and toxicology of repeated doses of hydrazine and 1,1-dimethylhydrazine in monkeys and rats, AMRL-TDR-64-43, *Aeromed. Res. Lab.*, Wright-Patterson Air Force Base, Ohio, 1964.
141. Peck, C. H., Report of the state botanist 1911, *N.Y. State Museum Bull.*, 157, 5, 1912.
142. Pilat, A., *Mushrooms*, H. W. Bijl, Amsterdam, 1954.
143. Pinkerton, M. K., Hagan, E. A., and Back, K. C., Distribution and excretion of ^{14}C-monomethylhydrazine, AMRL-TDR-67-175, *Aeromed. Res. Lab.*, Wright-Patterson Air Force Base, Ohio, 1967.
144. Prentiss, D. W., Five cases of mushroom-poisoning, three of which proved fatal; treatment of poisoning, *Philadelphia Med. J.*, 2, 607, 1898.
145. Puharich, A., *The Sacred Mushroom. Key to the Door of Eternity*, Doubleday, Garden City, N.Y., 1959.
146. Ramsbotton, J., Fungi and modern affairs, *Smithsonian Inst. Rep. 1945*, 313, 1945.
147. Ramsbottom, J., *Mushrooms and Toadstools*, New. Ed., Collins, London, 1954.
148. Rautavaara, T., *Suomen Sienstato*, Porwoo, Helsinki, 1947, (cf. Singer, 1954).
149. Reitz, R., *Mushroom Cookery*, Gramercy Publishing Co., New York, 1965.
150. Reynolds, B. A. and Thomas, A. A., Determination of hydrazine and 1-methylhydrazine in blood serum, AMRL-TDR-64-24, *Aeromed. Res. Lab.*, Wright-Patterson Air Force Base, Ohio, 1964.
151. Robinson, R. F. and Davidson, R. S., The large-scale growth of higher fungi, *Adv. Appl. Microbiol.*, 1, 261, 1959.
152. Rolfe, R. T. and Rolfe, F. W., *The Romance of the Fungus World*, Lippincott, Philadelphia, 1926.
153. Rombauer, I. M., *The Joy of Cooking*, Bobbs-Merrill, Indianapolis, 1936.
154. Ross, W., Ethnological notes on Mt. Hagen tribes, *Anthropos*, 31, 341, 1936.
155. Sapper, C., Pilzförmige Götzenbilder, *Globus*, 73, 327, 1898.
156. Schmidt, U., Grafen, P., and Goedde, H. W., Chemistry and biochemistry of α-lipoic acid, *Angew. Chem. [Engl.]*, 4(10), 846, 1965.
157. Schultes, R. E., The identification of teonanácatl, a narcotic Basidiomycete of the Aztecs, *Bot. Mus. Leafl. Harvard Univ.*, 5(5), 69, 1939.
158. Schultes, R. E., Teonanácatl, the narcotic mushroom of the Aztecs, *Am. Anthropol.*, N.S., 42, 429, 1940.
159. Schultes, R. E., The place of ethnobotany in the ethnopharmacologic search for psychomimetic drugs, *U.S. Public Health Serv. Publ.*, 1645, 33, 1967.
160. Scotti de Carolis, A., Lipparini, F., and Longo, V. G., Neuropharmacological investigations on muscimol, a psychotropic drug extracted from *Amanita muscaria, Psychopharmacologia*, 15, 186, 1969.
161. Seal, S. C., Gupta, N., and Kakoti, L. M., Mushroom poisoning in northeast Assam, *Indian M. Gaz.*, 86, 414, 1951.
162. Seaver, F. J., *The North American Cup-Fungi (Operculates)*, published by the author, New York, 1928.
163. Shaffer, R. L., Poisoning by *Pholiota squarrosa, Mycologia*, 57, 318, 1965.
164. Simandl, J. and Franc, J., Isolace tetraethylthiuramidsulfidu z hníku inko ustového, *Chem. Listy*, 50, 1862, 1956.
165. Simons, D. M., The mushroom toxins, *Del. Med. J.*, 43, 177, 1971.
166. Singer, R., Cryptogamic flora of the Arctic. VI. Fungi, *Econ. Bot.*, 20, 451, 1954.
167. Singer, R., Mycological investigations on teonanácatl, the Mexican hallucinogenic mushroom, I and II, *Mycologia*, 50, 239, 1958.

168. **Singer, R.**, A *Russula* provoking hysteria in New Guinea, *Mycopathol. Mycol. Appl.*, 9(4), 275, 1958.
169. **Singer, R.**, *Mushrooms and Truffles*, Leonard Hill (books) London, 1961.
170. **Singer, R. and Smith, A. H.**, New species of *Psilocybe*, *Mycologia*, 50, 141, 1958.
171. **Smith, A. H.**, *Mushrooms in Their Natural Habitats*, Vol. I, Sawyer's Inc., Portland, Ore., 1949.
172. **Smith, A. H.**, *The Mushroom Hunter's Field Guide*, University of Michigan Press, Ann Arbor, 1963.
173. **Smith, A. H. and Thiers, H. D.**, *The Boletes of Michigan*, University of Michigan Press, Ann Arbor, 1971.
174. **Strode, D. D.**, The Ocala deer herd, Fla. Game Freshwater Fish Comm. Publ. 1, Fed. Aid Proj. W-32 R, 1954.
175. **Taber, W. A.**, Fermentative production of hallucinogenic indole compounds, *Lloydia*, 30, 39, 1967.
176. **Theobald, W., Büch, O., Kunz, H. A., Krupp, P., Stenger, E. G., and Heimann, H.**, Pharmakologische und experimental-psychologische Untersuchungen mit 2 Inholtsstoffen Fliegenpilzes (*Amanita muscaria*), *Arzneim. Forsch.*, 18(3), 311, 1968.
177. **Thompson, J. E. S.**, An archeological reconnaissance in the Cotzumalhuapa region, Escuintla, Guatemala, Contribution to American anthropology and history, No. 44, *Carnegie Inst. Washington Publ.*, 574, 1948.
178. **Trotter, J. E.**, A report of nine cases of fungus poisoning, *Med. J. Aust.*, 1, 393, 1944.
179. **Tusa, R.**, The sport of gourmets, *Milwaukee Sentinel*, 3, 3, June 18, 1971.
180. **Tyler, V. E., Jr.**, Indole derivatives in certain North American mushrooms, *Lloydia*, 24, 71, 1961.
181. **Tyler, V. E., Jr., Benedict, R. G., Brady, L. R., and Robbers, J. E.**, Occurrence of amanita toxins in American collections of deadly amanitas, *J. Pharm. Sci.*, 55, 590, 1966.
182. **Tyler, V. E., Jr., Brady, L. R., Benedict, R. G., Khanna, J. M., and Malone, M. H.**, Chromatographic and pharmacologic evaluation of some toxic *Galerina* species, *Lloydia*, 26, 154, 1963.
183. **VanderVeer, J. B. and Farley, D. L.**, Mushroom poisoning (mycetismus), *Arch. Intern. Med.*, 55, 773, 1935.
184. **Vernet, J.**, Actions psichologique et therapeutique de la psilocybine, *Med. Hyg.*, 18, 420, 1960.
185. **Verrill, A. E.**, A recent case of mushroom intoxication, *Science*, N. S., 40(1029), 408, 1914.
186. **Vicuna, R., Philippi, F., and Melendez, O.**, Mushroom poisoning with case of familial incidence, *Rev. Med. Chile*, 72, 986, 1944.
187. **Waser, P. G.**, The pharmacology of *Amanita muscaria*, *U.S. Public Health Serv. Publ.*, 1645, 419, 1967.
188. **Wasson, R. G.**, Seeking the magic mushroom, *Life Magazine*, 42(19), 100, May 13, 1957.
189. **Wasson, R. G. and Wasson, V. P.**, *Mushrooms, Russia and History*, 2 vols., Pantheon Books, New York, 1957.
190. **Wasson, R. G.**, Wild mushrooms. A world of wonder and adventure, *The Herbarist*, 13, Boston, April, 1959.
191. **Wasson, R. G.**, Ololiuhqui and the other hallucinogens of Mexico, *Summa Anthropoligica en Homenaje a Roberto J. Weitlander*, 329, 1966.
192. **Wasson, R. G.**, Fly agaric and man, *U.S. Public Health Serv. Publ.*, 1645, 405, 1967.
193. **Wasson, R. G.**, *Soma, Divine Mushroom of Immortality*, Harcourt, Brace and World, New York, 1969.
194. **Wasson, R. G.**, Soma of the Aryans: an ancient hallucinogen?, *Bull. Narc.*, 22(3), 25, 1970.
195. **Wasson, R. G.**, The Soma of the RigVeda: what was it?, *J. Am. Oriental Soc.*, 91(2), 169, 1971.
196. **Webster, J. L. and Pickering, D.**, Acute mushroom poisoning: treatment with curare and intermittent positive pressure respiration, *Clin. Pediatr.*, 3, 523, 1964.
197. **Wieland, T.**, Poisonous principles of the genus *Amanita*, *Science*, 159, 946, 1968.
198. **Wieland, T. and Wieland, O.**, Chemistry and toxicology of *Amanita phalloides*, *Pharmacol. Rev.*, 11, 87, 1959.
199. **Wieland, T., Motzel, W., and Merz, H.**, Ueber das Vorkommen von Bufotenin im gelben Kollenblätterpilz, *Liebigs Ann.*, 581, 10, 1953.
200. **Wier, J. K. and Tyler, V. E.**, Investigation of *Coprinus atramentarius* for presence of disulfiram, *J. Am. Pharm. Assoc. (Sci. Ed.)*, 49, 429, 1960.
201. **Wilkinson, S.**, The history and chemistry of muscarine, *Q. Rev. Chem. Soc. London*, 15, 153, 1961.
202. **Wissoker, R.**, *From Generation to Generation*, Ad/Sell Co., Inc., St. Louis, Mo., 1971.
203. **Woolley, L.**, *History Unearthed*, Ernest Benn Ltd., London, 1963.
204. **Zeitlmayr, L.**, *Wild Mushrooms* (trans. from German), Frederick Muller, Ltd., London, 1968.

INDEX – PARTS I AND II

(p. 1–113, Part I; p. 115–209, Part II)

C

D

E

N

O

P

Q

R

S

T

U

V

ERRATUM, Part I

Page 43, right hand column, lines 3 and 4. *Change:* "Today most Roquefort and all its imitations are made from cow's milk." *to:* "Today most blue cheeses are made from cow's milk."